Symbian OS Explained

TITLES PUBLISHED BY SYMBIAN PRESS

- Symbian OS Explained
Jo Stichbury
0470 021306 416pp 2004 Paperback

- Symbian OS C++ for Mobile Phones, Volume 2
Richard Harrison
0470 871083 448pp 2004 Paperback

- Programming Java 2 Micro Edition on Symbian OS
Martin de Jode
0470 092238 498pp 2004 Paperback

- Symbian OS C++ for Mobile Phones, Volume 1
Richard Harrison
0470 856114 826pp 2003 Paperback

- Programming for the Series 60 Platform and Symbian OS
Digia
0470 849487 550pp 2002 Paperback

- Symbian OS Communications Programming
Michael J Jipping
0470 844302 418pp 2002 Paperback

- Wireless Java for Symbian Devices
Jonathan Allin
0471 486841 512pp 2001 Paperback

Symbian OS Explained

Effective C++ Programming for Smartphones

Jo Stichbury

Reviewed by

David Batchelor, Andy Cloke, Reem El Ghazzawi, Martin Hardman, Morgan Henry, John Pagonis, William Roberts, Keith Robertson, Phil Spencer, Colin Turfus

Managing editor

Phil Northam

Project editor

Freddie Gjertsen

John Wiley & Sons, Ltd

Transferred to digital printing, 2005

Email (for orders and customer service enquiries): cs-books@wiley.co.uk
Visit our Home Page on www.wileyeurope.com or www.wiley.com

Other Wiley Editorial Offices

John Wiley & Sons Inc., 111 River Street, Hoboken, NJ 07030, USA

Jossey-Bass, 989 Market Street, San Francisco, CA 94103-1741, USA

Wiley-VCH Verlag GmbH, Boschstr. 12, D-69469 Weinheim, Germany

John Wiley & Sons Australia Ltd, 33 Park Road, Milton, Queensland 4064, Australia

John Wiley & Sons (Asia) Pte Ltd, 2 Clementi Loop #02-01, Jin Xing Distripark, Singapore 129809

John Wiley & Sons Canada Ltd, 22 Worcester Road, Etobicoke, Ontario,
Canada M9W 1L1

Wiley also publishes its books in a variety of electronic formats. Some content that appears in print may not be available in electronic books.

Library of Congress Cataloging-in-Publication Data

Stichbury, Jo.
Symbian OS explained effective C++ programming for smartphones / By Jo Stichbury.
p. cm.
Includes bibliographical references and index.
ISBN 0-470-02130-6 (pbk. alk. paper)
1. Mobile communication systems–Computer programs. 2. Operating systems (Computers) 3. C++ (Computer program language) I. Title.

TK6570.M6S745 2004
005.265–dc22 2004015414

British Library Cataloguing in Publication Data

A catalogue record for this book is available from the British Library

ISBN 0-470-02130-6

Typeset in 10/12pt Optima by Laserwords Private Limited, Chennai, India

This book is printed on acid-free paper responsibly manufactured from sustainable forestry in which at least two trees are planted for each one used for paper production.

Contents

Foreword

Charles Davies, Chief Technical Officer, Symbian

Software engineers live in interesting times: software is becoming pervasive. We all increasingly rely on our personal computers and use their software as an essential tool in organizing our lives. But what we "see" is just the tip of the iceberg. Most software exists beneath the surface within a variety of embedded systems such as electronic consumer devices, motor cars, and aircraft. Symbian OS is targeted at mobile phones – a class of embedded system that exists in massive volume and which is used by the entire developed world.

The amount of software built into a mobile phone is expanding rapidly. In recent times it has outpaced Moore's law: in the past three years the amount of embedded software in high-end phones has jumped from about 2 MB to 20 MB. This is partly required by the sophistication of new 3G networks, but it is mainly due to mobile phones subsuming the functionality of other portable consumer devices, such as digital cameras and camcorders, digital audio players, video players, electronic organizers, mobile gaming consoles, portable radios, portable TVs, email terminals, cordless phones and even electronic payment cards. The mobile phone is becoming *the* key portable lifestyle support system – an electronic Swiss army knife.

Symbian OS is needed because this explosion in software-supported functionality requires a capable operating system that is designed for sophisticated, always-on, battery-powered mobile devices.

The object-oriented programming paradigm of Symbian OS helps manage system complexity and permeates the architecture of Symbian OS. This architecture uses many advanced, but classical, constructs found in other multitasking operating systems. Examples include pre-emptive multitasking threads, processes, asynchronous services and internal servers for serializing access to shared resources. Symbian OS has some particular features that also need to be understood if one is to become an effective Symbian OS programmer. These distinct features have been designed to cope with the rigorous discipline of mobile device programming, for example, in the handling of asynchronous events and errors, avoiding memory leakages and other dangling resources.

Software engineers moving from the embedded space will probably have to make a transition from C to the object-oriented world of C++. Software engineers moving from the PC space will probably be used to C++, but will not be used to the tougher disciplines of programming for mobile phones where robustness, code size, memory usage, performance and battery life is important, where there are no regular reboots to mop up memory leaks and where the radio increases and enriches the number of events to which applications have to respond.

Whatever your background and level of experience, your effectiveness will be much improved if you read this book and understand the essential Symbian OS concepts.

About this Book

Developing good C++ code on Symbian OS requires a clear understanding of the underlying concepts and the nature of the operating system. This book explains the key features of Symbian OS and shows how you can use this knowledge most effectively. It also focuses on some aspects of good C++ style that particularly apply to Symbian OS. With understanding and practice, the expertise required to write high quality C++ code on Symbian OS should become second nature.

The book is divided into a series of self-contained chapters, each discussing specific and important features of Symbian OS. Besides the fundamentals, the chapters illustrate best practice and describe any common mistakes to avoid. The chapters are concise enough to impart the insight you need without being so brief as to leave you needing more information. Each chapter delivers simple and straightforward explanations without omitting the important facts.

This book doesn't teach you how to write C++. It assumes you're already familiar with the most important concepts of the language. Neither does it walk through Symbian OS from the perspective of developing a particular application. Instead, I try to impart an understanding of the central concepts and key features of Symbian OS and good techniques in C++. To quote from Scott Meyers,[1] whose book inspired me to write this one, "In this book you'll find advice on what you should do, and why, and what you should not do, and why not".

---

[1] Scott Meyers, *Effective C++: 50 specific ways to improve your programs and designs*, 1997. See the Bibliography for further details.

Who Is It For?

The book assumes a reasonable understanding of programming in C++. It does not assume in-depth knowledge of Symbian OS and covers the basics such as descriptors (Chapters 5 and 6) and active objects (Chapters 8 and 9) as well as more complex features such as the Symbian OS client–server architecture (Chapters 11 and 12) and ECOM (Chapter 14).

The book focuses upon the core of the operating system, which should be equally applicable to all versions of Symbian OS, and user interfaces, such as UIQ and Series 60. If there are any important differences between platforms, they are highlighted. At the time of writing (Spring 2004), Symbian is preparing to release a new version of the operating system, Symbian OS v8.0. This book explicitly indicates any changes that the new version will introduce, where the author is aware of them.

If you are a developer targeting, or thinking of targeting, Symbian OS, this book will show you how to write the most effective C++. You will benefit from a greater understanding of the characteristic features and design of the operating system, and confidence in how to use them. However experienced you are with Symbian OS, there are always new tricks to learn, which is why this book will appeal to all levels of developer. It reflects the combined wisdom of the many experienced Symbian OS developers I have worked with. Over the years, they have taught me a great deal, and when I looked more closely at parts of the OS whilst writing this book, I learnt even more. I hope you will too.

How to Use This Book

As I've already mentioned, the book is split into a number of chapters, where each chapter acts as a guide to a particular feature of Symbian OS. The title of each, a detailed table of contents and the index, glossary and bibliography sections are all designed to help you find the information you need.

The chapters do not necessarily need to be read in sequence. Rather, the book can be dipped into for revision, reference or to provide a handy "tip for the day"; it does not necessarily need to be read from cover to cover. The chapters cross-reference each other where there is overlap between them, to indicate which other areas are particularly relevant to the discussion.

For clarification and explanation, the chapters also use example code which has been designed to illustrate good coding style and the conventions of Symbian OS code and code layout.

Notation and Code Conventions Used in This Book

The textual layout of this book is self-explanatory, but the code layout needs some introduction. Where I use example code, it will be highlighted as follows:

```
This is example code;
```

C++ code for Symbian OS uses an established naming convention which you should try to stick to in order for your own code to communicate its intent most clearly. Besides communication, the main benefit of adhering to the convention is that it is chosen to reflect clearly object cleanup and ownership.

The best way to get used to it is to look at code examples such as those found in your chosen SDK as well as in this book. The main features of the naming conventions are described here; in Chapter 1, I discuss the class name prefix conventions, while Chapter 2 covers the reason for the trailing L on some function names and Chapter 3 discusses the use of trailing C.

Capitalization

The first letter of class names should be capitalized:

```
class TClanger;
```

The words making up variable, class or function names should be adjoining, with the first letter of each word capitalized where appropriate (parameters, automatic, global and member variables, and function parameters have a lower case first letter). The rest of each word should be lower case, including acronyms:

```
void SortFunction();
TInt myLocalVariable;
CMemberVariable* iDevilsHaircut;
class CActiveScheduler;
class CBbc;//Acronyms are not usually written in upper case
```

Global variables tend to be discouraged, but typically start either with an initial capital letter or are prefixed with a lower case "g".

Prefixes

Member variables are prefixed with a lower case "i" which stands for "instance".

```
TInt iCount;
CPhilosopher* iThinker;
```

Parameters are prefixed with a lower case "a" which stands for "argument". Do not use "an" for arguments that start with a vowel.

```
void ExampleFunction(TBool aExampleBool, const TDesC& aName);
```

Note `TBool aExampleBool` rather than `TBool anExampleBool`.
Automatic variables have no prefix and the first letter is lower case.

```
TInt index;
CMyClass* ptr = NULL;
```

Class names should be prefixed with an appropriate letter ("C", "R", "T" or "M" as described fully in Chapter 1).

```
class CActive;
class TParse;
```

Constants should be prefixed with "K".

```
const TInt KMaxFilenameLength = 256;
#define KMaxFilenameLength 256
```

Enumeration members are prefixed with "E". Enumerations are types, and so are prefixed with "T" (you can find more information about T classes in Chapter 1).

```
enum TChilliPeppers {EScotchBonnet, EJalapeno, ECayenne};
```

Suffixes

A trailing "L" on a function name indicates that the function may leave.

```
void ConstructL();
```

A trailing "C" on a function name indicates that the function returns a pointer that has been pushed onto the cleanup stack.

```
CPhilosopher* NewLC();
```

A trailing "D" on a function name means that it will result in the deletion of the object referred to by the function.

```
TInt ExecuteLD(TInt aResourceId);
```

Underscores

Underscores should be avoided except in macros (`__ASSERT_DEBUG`) or resource files (`MENU_ITEM`).

Code layout

You'll notice that the curly bracket layout in Symbian OS code, used throughout this book, is to indent the bracket as well as the following statement. I'm not a big fan of this, and I don't think it makes that much difference if you use an alternative layout. However, if you want to stick to writing code the "Symbian way", you should adhere to the following convention:

```
void EatChilliL(CChilliPepper* aChilli)
    {
    if (!aChilli)
        {
        User::Leave(KErrArgument);
        }
    TChilliType type = aChilli->Type();
    if (EScotchBonnet==type)
        {
        User::Leave(KErrNotSupported);
        }
    DoEatChilliL(aChilli);
    }
```

Tips

Throughout the book, tips and advice are signified by a symbol in the margin.

These tips complement each section, providing reminders of key elements of the discussion.

Introduction to Symbian OS

A key characteristic of a mobile phone is that it is small – as small and as light as possible.

It's with you all the time, ready to send and receive phone calls or messages, play alarms to wake you, connect to the phone network or other devices, organize your personal information or play games. Whatever you want to do, you want access to your phone to be instant. It should be ready without the long boot-up you expect from a PC. It should react to user input from the moment you pick it up. And it should be reliable. As a mobile phone owner, how many of your contacts' phone numbers do you know? Your phone keeps your personal data for you and it's essential that it doesn't lose it.

Let's examine the consequences of these features on the design of a mobile operating system, such as Symbian OS. The phone may be small and light, but, as users, we still demand it to have a reasonable battery life. This places huge importance on efficient power management. The operating system cannot drain the battery and must allow the processor to power parts of the system down where possible, although it cannot ever power off completely because it must handle incoming calls and messages, and signal alarms.

The user expects the phone to be responsive, not sluggish to respond to each key press; the operating system and hardware must carefully balance demands for good performance speed with the consumption requirements of power-hungry processors. Costs are also important: they limit the processor and amount of memory in a mobile device. The operating system must be efficient, to make best use of the limited processor and memory resources available, whilst using the least power.

Besides being efficient, the operating system must be robust when the limited resources are exhausted. It must be engineered to cope with low memory conditions, loss of power or when a communications link is unavailable. Memory management is key. The operating system must track precious system resources accurately and free them when they are not required. It's not acceptable for memory to slowly leak away, resulting in disintegration in performance and usability until the user is forced to reboot. The operating system should make it easy for software engineers to write code that runs without memory leaks and can handle out-of-memory conditions when they occur.

The mobile phone market is a mass market, with many millions of units shipped. It's difficult, if not impossible, to recall them or require the user to upgrade through service packs. So when a phone is shipped, it's shipped. It must not have any serious defects. Not only must the platform on which a phone is based be well-engineered, it must also provide the means for developers to build, debug and test robust code.

As users, we also demand that our mobile phones are as cheap as possible. We want the latest, trendiest phone, which may just be smaller than the last, or have additional features like Bluetooth, an integrated camera or video player, or an MP3 player. This means that the lifetime of a phone in the marketplace is limited. When a manufacturer develops a phone it must be ready for market as quickly as possible and the operating system should ideally be flexible so the basic design can be extended or upgraded and released in a different model.

Symbian OS was designed for mobile devices, from its earliest incarnation as EPOC32 in the Psion Series 5. Many of the requirements of today's mobile phones applied equally well in those days, and its design reflects that. The huge demands upon a mobile operating system have shaped Symbian OS, from resilient power-management and careful use of memory resources, to the sophisticated use of C++ and object-oriented programming techniques. As a developer on Symbian OS, you benefit from a platform which was created specifically for mobile devices and which has evolved with the market.

Of course, the responsibilities don't stop with the operating system. To work most successfully on a mobile phone, your code must also be efficient, robust, responsive and extensible. Some of the problems of mobile computing are notoriously difficult to solve and programming in C++ is complex at the best of times. It's not easy, but opting to work on Symbian OS gives you the benefits of a purpose-built platform. This book brings you the best advice from those experienced in working with it.

Author Biography

Jo Stichbury

Jo Stichbury was educated at Magdalene College, Cambridge, where she held the Stothert Bye-Fellowship. She has an MA in Natural Sciences (Chemistry) and a PhD in the chemistry of organometallic Molybdenum complexes. After a brief spell in postdoctoral research at Imperial College, she joined Psion Software in 1997, when Symbian OS was still known fondly as EPOC32. She has worked with the operating system ever since, within the Base, Connectivity and Security teams of Symbian, and also for Advansys, Sony Ericsson and Nokia.

As the contents of this book will reveal, Jo has a somewhat unhealthy interest in the Clangers and Greek mythology. She currently lives in Vancouver with her partner and two cats.

Author's Acknowledgments

Once upon a time, when I joined Psion Software, there weren't any books available on Symbian OS (which was known then as EPOC32). So I'd like to acknowledge, with gratitude, the support I received from Morgan Henry, Matthew Lewis, Peter Scobie and Graham Darnell, who helped me take my first steps as a programmer. I intend this book to record their insight and knowledge as I've interpreted it, as well as that of all the other Symbian OS developers I've worked with since.

I am grateful to everyone who has contributed to this book. Leon Clarke, Will Bamberg, Mark Jacobs and Paul Stevens made useful suggestions at the early stages; later on, Julian Lee, Keith Robertson and Dennis May answered my technical questions. Will Bamberg provided me with his own well-crafted documents on leaves and out-of-memory testing – and didn't complain when I de-articulated them for inclusion in Chapters 2 and 17.

The input from all my reviewers was much appreciated. I would like to thank the official reviewers: Keith, David, John, Colin, Andrew, Morgan, Martin, Phil and Reem who all contributed enormously to the technical accuracy of this book and to my peace of mind. I also received many helpful contributions from Peter van Sebille and John Blaiklock at Sony Ericsson, Mark Jacobs and William Roberts at Symbian and Will Bamberg.

I'm very grateful to Dave Morten, Steve Burke and Brigid Mullally at Sony Ericsson, for showing understanding about the time it took me to write this book. I couldn't have done it without their cooperation. The diagrams in this book were inspired by Brigid.

This book would also have been impossible without the help and support of my partner Mark, who makes everything possible. Since I started writing, I've also acquired two Siamese cats, who have spent more time distracting me than helping but, for that alone, they deserve inclusion here.

I would like to thank Symbian Press, particularly Freddie Gjertsen and Phil Northam, for their patience and fortitude. Thanks must also go to Gaynor Redvers-Mutton at John Wiley and to Karen Mosman who started the ball rolling.

I'd also like to thank Oliver Postgate for granting me permission to include the Clangers.

Symbian Press Acknowledgments

First and foremost we would like to thank Jo for her tireless endeavour in the production of this book.

Thanks are also due to all our reviewers, named and otherwise, and especially to young Phil, choleric beyond his years. Where would we be without "constructive" criticism?

And last, but not least, thanks to the lovely Boundary Row tea-ladies, Gayle and Victoria. Without constant refreshment to sustain us, Symbian Press would be nought.

Cover design based upon a concept by Jonathan Tastard.

Code checklist based on the documentation of System Management Group, Symbian.

1

Class Name Conventions on Symbian OS

Llanfairpwllgwyngyllgogerychwyrndrobwllllantysiliogogogoch

The longest place name in the British Isles, said to be invented to attract tourists

Symbian OS defines several different class types, each of which has different characteristics. The categories are used to describe the main properties and behavior of objects of each class, such as where they may be created (on the heap, on the stack or on either) and, particularly, how they should be cleaned up. Each of the class types has a well-defined set of rules which makes the creation and destruction of instances of that class quite straightforward.

To enable the types to be easily distinguished, Symbian OS uses a simple naming convention which prefixes the class name with a letter (usually T, C, R or M). Naming conventions aren't always popular, but this one isn't difficult to follow and is clearly of value, since it allows the behavior of a class to be easily identified, particularly with respect to cleanup. As a class designer, the class types simplify matters. You consider the required behavior of your class, and match it to the definitions of the basic Symbian OS types. Having chosen a type, you can then concentrate on the role of the class. By the same token, as a client of an unfamiliar class, the naming convention indicates how you are expected to instantiate an object, use it and then destroy it in a leave-safe way.

1.1 Fundamental Types

I'll discuss the main features of each class type in this chapter. However, before doing so let's go back to basics and consider the fundamental types. Symbian OS provides a set of `typedef`s of the built-in types, which are guaranteed to be compiler-independent; these should always be used instead of the native types.

- `TIntX` and `TUintX` (for `X` = 8, 16 and 32) for 8-, 16- and 32-bit signed and unsigned integers respectively. Unless you have a good reason to do so, such as for size optimization or compatibility, you should use the non-specific `TInt` or `TUint` types, which correspond to signed and unsigned 32-bit integers, respectively.
- `TInt64`. Releases of Symbian OS prior to v8.0 had no built-in ARM support for 64-bit arithmetic, so the `TInt64` class implemented a 64-bit integer as two 32-bit values. On Symbian OS v8.0, `TInt64` and `TUInt64` are `typedef'd` to `long long` and use the available native 64-bit support.
- `TReal32 and TReal64` (and `TReal`, which equates to `TReal64`) for single- and double-precision floating point numbers, equivalent to `float` and `double` respectively.[1] Operations on these are likely to be slower than upon integers so you should try to avoid using them unless necessary.
- `TTextX` (for `X` = 8 or 16), for narrow and wide characters, correspond to 8-bit and 16-bit unsigned integers, respectively.
- `TAny*` should be used in preference to `void*`, effectively replacing it with a `typedef'd` "pointer to anything". `TAny` is thus equivalent to `void` but, in the context where `void` means "nothing", it is not necessary to replace the native `void` type. Thus, a function taking a `void*` pointer (to anything) and returning `void` (nothing) will typically have a signature as follows on Symbian OS:

```
void TypicalFunction(TAny* aPointerParameter);
```

  This is the one exception to the rule of replacing a native type with a Symbian OS `typedef`; it occurs because `void` is effectively compiler-independent when referring to "nothing".

- `TBool` should be used for boolean types. For historical reasons `TBool` is equivalent to `int` and you should use the Symbian OS `typedef'd` values of `ETrue` (=1) and `EFalse` (=0). Bear in mind that C++ will interpret any nonzero value as true. For this reason, you should refrain from making direct comparisons with `ETrue`.
- Each `TBool` represents 32 bits, which may be quite wasteful of memory in classes with a number of flags representing state or settings.

[1] Note that these are `typedefs` and should not be confused with the Symbian OS `TRealX` class, which describes an extended 64-bit precision real value.

You may wish to use a bitfield rather than a number of booleans, given that the 32 bits of a single `TBool` could hold 32 boolean values in a bitfield. Of course, you should consider the potential code complexity, and trade that off against the benefits of a smaller object.

The `typedef`'d set of Symbian OS built-in types are guaranteed to be compiler-independent and should be used instead of the native types except when returning `void` which is equivalent to "nothing".

1.2 T Classes

T classes behave much like the C++ built-in types, hence they are prefixed with the same letter as the `typedefs` described above (the "T" is for "Type"). Like built-in types they have no destructor and, consequently, T classes should not contain any member data which itself has a destructor. Thus, a T class will contain member data which is either:

- "plain ol' data" (built-in types) and objects of other T classes
- pointers and references with a "uses a" relationship rather than a "has a" relationship, which implies ownership. A good example of this is the `TPtrC` descriptor class, described in Chapter 5.

T classes contain all their data internally and have no pointers, references or handles to data they own (although references to data owned by other objects is allowed). The reason for not allowing ownership of external data is because the T class must not have a destructor.

Without a destructor, an object of a T class can be created on the stack and will be cleaned up correctly when the scope of that function exits, either through a normal return or a leave ("leaving" is discussed in detail in Chapter 2). If a T class were to have a destructor, Symbian OS would not call it in the event of a leave because leaves do not emulate the standard C++ `throw` semantics. If a call to the destructor were necessary for the object to be safely cleaned up, the object could only be created on the stack in the scope of code which can be guaranteed not to leave – which is clearly somewhat restrictive.

An object of a T class can also be created on the heap. Such an object should be pushed onto the cleanup stack prior to calling code with the potential to leave. In the event of a leave, the memory for the T object is deallocated by the cleanup stack (which is discussed in detail in Chapter 3) but no destructor call is made.

T classes are also often defined without default constructors; indeed, if a T class consists only of built-in types, a constructor would prevent member initialization as follows:

```
TMyPODClass local = {2000, 2001, 2003};
```

However, in the rare case that the T class has exported virtual functions, a default constructor must be exported because it is required for any client code to link against. The reasons for this are discussed in Chapter 20, which explains the `EXPORT_C` syntax.

As a rule, the members of a T class will be simple enough to be bitwise copied, so copy constructors and assignment operators are trivial and the implicit compiler-generated versions are likely to be more efficient. So you generally don't need to write a copy constructor or assignment operator unless, of course, you want to prevent cloning, in which case you should declare them both private to your class and leave them unimplemented.

Some T classes have fairly complex APIs, such as `TLex`, for lexical analysis, and the descriptor base classes `TDesC` and `TDes`, described in Chapter 5. In other cases, a T class is simply a C-style `struct` consisting only of public data (traditionally, a `struct` was prefixed with S instead of T, but more recent Symbian OS code tends to make these T classes too).

You'll find the T prefix used for enumerations too, since these are simple types. For example:

```
enum TMonthsOfYear {EJanuary = 1, EFebruary = 2, ..., EDecember = 12};
```

A T class must not have a destructor.

1.3 C Classes

Classes with the C prefix[2] ultimately derive from class `CBase` (defined in `e32base.h`). This class has two characteristics which are inherited by its subtypes and thus guaranteed for every C class.

Firstly, `CBase` has a virtual destructor so a `CBase`-derived object may be destroyed properly by deletion through a `CBase` pointer. This

[2] In case you are wondering, the "C" stands for "Class", which perhaps makes a "C class" something of a tautology, though it is an accurate reflection of the fact that it is more than the simple "Type" described by a T class.

is common when using the cleanup stack, since the function invoked when pushing a CBase-derived object onto the cleanup stack is the CCleanupStack::PushL(CBase* aPtr) overload.

If CCleanupStack::PopAndDestroy() is called for that object (or if a leave occurs), the object is deleted through the CBase pointer. The virtual destructor in CBase ensures that C++ calls the destructors of the derived class(es) in the correct order (starting from the most derived class and calling up the inheritance hierarchy). So, as you'll have gathered, unlike T classes, it's fine for a C class to have a destructor, and they usually do.

It is worth noting at this point that, if the class does not derive from CBase and is pushed onto the cleanup stack, the CCleanupStack::PushL(TAny* aPtr) overload will be used instead. As described above, when PopAndDestroy() is called, or a leave occurs, the memory for the object will be deallocated but no destructor calls made. So if you do not inherit from CBase, directly or indirectly, even if your base class destructor has a virtual destructor, objects of your class will not be cleaned up as you expect.

The second characteristic of CBase, and its derived classes, is that it overloads operator new to zero-initialize an object when it is first allocated on the heap. This means that all member data in a CBase-derived object will be zero-filled when it is first created. You don't have to do this explicitly yourself in the constructor. Zero initialization will not occur for stack objects because allocation on the stack does not use operator new. This could potentially cause unexpected or different behavior between zero-filled heap-based and non-zeroed stack-based objects. For this reason, among others such as managing cleanup in the event of a leave, objects of a C class must *always* be allocated on the heap.

Clearly, when it is no longer needed, a heap-based object must be destroyed. Objects of C classes typically exist either as pointer members of another class or are accessed by local pointer variables. If owned, the CBase-derived object should be destroyed by a call to delete, for example in the destructor of the owning class. If the C class object is not owned, instead being accessed through a local pointer, that pointer must be placed on the cleanup stack prior to calling any code with the potential to leave – otherwise it will be orphaned on the heap in the event of a leave. I'll discuss this further in Chapter 3.

If you look at e32base.h, you'll notice that CBase declares a private copy constructor and assignment operator. This is a common strategy used to prevent a client from making accidental shallow copies of, or assignments to, objects of a class. If such an operation is valid for your class you must explicitly declare and define a public copy constructor and assignment operator, because their private declaration in the base class means they cannot be called implicitly. However, given the nature of C classes, a deep copy operation may well have the potential to leave, and you should never allow a constructor (or destructor) to leave (I describe the reasons for this in more detail in Chapter 4). If you need to allow C class copying, rather than defining and implementing a public copy constructor, you should add a leaving function, e.g. `CloneL()` or `CopyL()`, to your class to perform the same role.

Since most C classes tend not to be straightforward enough to be bitwise copied, the implicit copy is best avoided, and this is yet another benefit of deriving from `CBase`. The private declaration of a copy constructor and assignment operator in the `CBase` class means you don't have to privately declare them in every C class you write in order to prevent clients from making potentially unsafe "shallow" copies.

Objects of a C class must always be allocated on the heap.

1.4 R Classes

The "R" which prefixes an R class indicates a resource, usually a handle to an external resource such as a session with the file server. There is no equivalent `RBase` class, so a typical R class will have a constructor to set its resource handle to zero, indicating that no resource is currently associated with the newly constructed object. You should not attempt to initialize the resource handle in the constructor, since it may fail; you should never leave in a constructor, as I'll explain in Chapter 4.

To associate the R class object with a resource, the R class will typically have a function such as `Open()`, `Create()` or `Initialize()` which will allocate the resource and set the handle member variable as appropriate or fail, returning an error code or leaving. The class will also have a corresponding `Close()` or `Reset()` method, which releases the resource and resets the handle value to indicate that no resource is associated with the object. It should thus be safe to call such a function more than once on the same object. Although, in theory, the cleanup function can be named anything, by convention it is almost always called `Close()`.

A common mistake when using R classes is to forget to call `Close()` or to assume that there is a destructor which cleans up the owned resource. This can lead to serious memory leaks.

R classes are often small, and usually contain no other member data besides the resource handle. It is rare for an R class to have a destructor – it generally does not need one because cleanup is performed in the `Close()` method.

R classes may exist as class members or as automatic variables on the stack, or occasionally on the heap. You must ensure the resource will be released in the event of a leave, typically by using the cleanup stack, as described in Chapter 3. Remember, if using an R class object as a heap-based automatic variable, you must make sure that the resource is released as well as freeing the memory associated with the object itself, typically by using two push calls: `CleanupClosePushL()`, or a similar function, to ensure that the resource is cleaned up, and a standard `CleanupStack::PushL(TAny*)` which simply calls `User::Free()` on the heap cell.

The member data of an R class is typically straightforward enough to be bitwise copied, so you would not expect to see an explicit copy constructor or assignment operator unless a shallow copy would result in undefined behavior. An example of this might be where the duplication of a handle value by bitwise copy would make it ambiguous as to which object owns the resource. In these circumstances undefined behavior might result if both copies attempt to release it. Calling the `Close()` function on the same object may be safe because the handle value member is reset, but the same cannot always be said for calling `Close()` twice on the same underlying resource via two separate handle objects. If one object frees the resource, the handle held by the other is invalid.

If your class contains a handle to a resource which cannot be shared safely by bitwise copy, you should declare a cloning function in the event that it must be copied. If you want to prevent any kind of copying for objects of your R class, you should declare, but not define, a private copy constructor and assignment operator.

The rules for R classes are more open to interpretation than C or T classes, so you'll find more different "species" of R class. Within Symbian OS the types of resources owned by R classes vary from ownership of a file server session (class `RFs`) to ownership of memory allocated on the heap (class `RArray`).

`Close()` must be called on an R class object to cleanup its resource.

1.5 M Classes

Computing folklore relates that "mix-ins" originated from Symbolic's Flavors, an early object-oriented programming system. The designers were apparently inspired by Steve's Ice Cream Parlor, a favorite ice cream shop of MIT students, where customers selected a flavor of ice cream (vanilla, strawberry, chocolate, etc) and added any combination of mix-ins (nuts, fudge, chocolate chips and so on).

When referring to multiple inheritance, it implies inheriting from a main "flavor" base class with a combination of additional "mix-in" classes which extend its functionality. I believe the designers of Symbian OS adopted the term "mixin", and hence the M prefix in turn, although the use of multiple inheritance and mixins on Symbian OS should be more controlled than a trip to an ice cream parlor.

In general terms, an M class is an abstract interface class. A concrete class deriving from such a class typically inherits a "flavor" base class (such as a `CBase`, or a `CBase`-derived class) as its first base class, and one or more M class "mixin" interfaces, implementing the interface functions. On Symbian OS, M classes are often used to define callback interfaces or observer classes.

An M class may be inherited by other M classes. I've shown two examples below. The first illustrates a concrete class which derives from `CBase` and a single mixin, `MDomesticAnimal`, implementing the functions of that interface. The `MDomesticAnimal` class itself derives from, but does not implement, `MAnimal`; it is, in effect, a specialization of that interface.

```
class MAnimal
    {
public:
    virtual void EatL() =0;
    };

class MDomesticAnimal : public MAnimal
    {
public:
    virtual void NameL() =0;
    };

class CCat : public CBase, public MDomesticAnimal
    {
public:
    virtual void EatL();  // From MAnimal, via MDomesticAnimal
    virtual void NameL(); // Inherited from MDomesticAnimal
    ...                   // Other functions omitted for clarity
    };
```

The second example illustrates a class which inherits from `CBase` and two mixin interfaces, `MRadio` and `MClock`. In this case, `MClock` is not a specialization of `MRadio` and the concrete class instead inherits from them separately and implements both interfaces. For M class inheritance, various combinations of this sort are possible.

```
class MRadio
    {
public:
    virtual void TuneL() =0;
    };
```

```
class MClock
    {
public:
    virtual void CurrentTimeL(TTime& aTime) =0;
    };

class CClockRadio : public CBase, public MRadio, public MClock
    {
public:
    virtual void TuneL();
    virtual void CurrentTimeL(TTime& aTime);
    ... // Other functions omitted for clarity
    };
```

The use of multiple *interface* inheritance, as shown in the previous examples, is the *only* form of multiple inheritance encouraged on Symbian OS. Other forms of multiple inheritance can introduce significant levels of complexity, and the standard base classes were not designed with it in mind. For example, multiple inheritance from two `CBase`-derived classes will cause problems of ambiguity at compile time, which can only be solved by using virtual inheritance:

```
class CClass1 : public CBase
{...};

class CClass2 : public CBase
{...};

class CDerived : public CClass1, public CClass2
{...};

void TestMultipleInheritance()
    {
    // Does not compile, CDerived::new is ambiguous
    // Should it call CBase::new from CClass1 or CClass2?
    CDerived* derived = new (ELeave) CDerived();
    }
```

Let's consider the characteristics of M classes, which can be thought of as equivalent to Java `interfaces`. Like Java `interface`, an M class should have no member data; since an object of an M class is never instantiated and has no member data, there is no need for an M class to have a constructor.

In general, you should also consider carefully whether an M class should have a destructor (virtual or otherwise). A destructor places a restriction on how the mixin is mixed in, forcing it to be implemented only by a `CBase`-derived class. This is because a destructor means that `delete` will be called, which in turn demands that the object cannot reside on the stack, and must always be heap-based. This implies that an implementing class must derive from `CBase`, since T classes never possess destructors and R classes do so only rarely.

In cases where the ownership of an object that implements an M class is through a pointer to that interface class, it is necessary to provide some means of destroying the object. If you know you *can* restrict implementers of your interface class to derivation from `CBase`, then you can provide a virtual destructor in the M class. This allows the owner of the M class pointer to call `delete` on it.

If you do not define a virtual destructor, deletion through an M class pointer results in a `USER 42` panic, the reasons for which are as follows: a concrete class that derives from an M class also inherits from another class, for example `CBase`, or a derived class thereof. The mixin pointer should be the second class in the inheritance declaration order[3], and the M class pointer will thus have been cast to point at the M class subobject which is some way into the allocated heap cell for the object, rather than at the start of a valid heap cell.

The memory management code attempting to deallocate the heap space, `User::Free()`, will panic if it cannot locate the appropriate structure it needs to cleanup the cell. This is resolved by using a virtual destructor.

However, as an alternative to defining the interface as an M class, you should consider simply defining it as an abstract `CBase`-derived class instead, which can be inherited as usual by a C class. `CBase` already provides a virtual destructor, and nothing else, so is ideal for defining an interface class where its implementation classes can be limited to C classes. Of course, in these cases, the implementation classes will be limited to single inheritance, because, as I described above, multiple inheritance from `CBase` results in ambiguity and the dreaded "diamond-shape" inheritance hierarchy.

In general, a mixin interface class should not be concerned with the implementation details of ownership. If it is likely that callers will own an object through a pointer to the M class interface, as described above, you must certainly provide a means for the owner to relinquish it when it is no longer needed. However, this need not be limited to cleaning up through a destructor. You may instead provide a pure virtual `Release()` method so the owner can say "I'm done" – it's up to the implementing code to decide what this means (for a C class, the function can just call "`delete this`"). This is a more flexible interface – the implementation class can be stack- or heap-based, perform reference

[3] The correct class definition is

```
class CCat : public CBase, public MDomesticAnimal{...};
```

and not

```
class CCat : public MDomesticAnimal, public CBase{...};
```

The "flavor" C class must always be the first specified class of the base class list, to emphasize the primary inheritance tree. It also enables C class objects of derived classes such as these to be placed on the cleanup stack using the correct `CleanupStack::PushL()` overload (see Chapter 3 for more details).

counting, special cleanup or whatever. By the way, it isn't essential to call your cleanup method `Release()` or `Close()`, but it can help your callers if you do. First of all, it's recognizable and easy enough to guess what it does. More importantly, it enables the client to make use of the `CleanupReleasePushL()` function described in Chapter 3.

Like a Java `interface`, an M class should usually have only pure virtual functions. However, there may be cases where non-pure virtual functions may be appropriate. A good example of this occurs when all the implementation classes of that interface have common default behavior. Adding a shared implementation to the interface reduces code duplication and its related bloat and maintenance headaches. Of course, there's a restriction on what this default implementation can do, because the mixin class must have no member data. Typically, all virtual functions are implemented in terms of calls to the pure virtual functions of the interface.

An M class has similar characteristics to a Java `interface`. It has no member data and no constructor. The use of multiple interface inheritance using M classes is the only form of multiple inheritance encouraged on Symbian OS.

1.6 Static Classes

We've reached the end of the naming convention prefixes, though not quite the end of the types of class commonly found on Symbian OS. There are a number of classes, with no prefix letter, that provide utility code through a set of static member functions, for example, `User` and `Mem`. The classes themselves cannot be instantiated; their functions must instead be called using the scope resolution operator:

```
User::After(1000); // Suspends the current thread for 1000 microseconds
Mem::FillZ(&targetData, 12); // Zero-fills the 12-byte block starting
                             // from &targetData
```

Classes which contain only static functions do not need to have a name prefix.

1.7 Buyer Beware

It's a good rule of thumb that for all rules there are exceptions. The class name conventions on Symbian OS are no exception to that rule and there are classes in Symbian OS code itself which do not fit the ideals I've put to you above. There are a few classes in Symbian OS which don't even conform to the naming conventions. Two well-documented exceptions are the kernel-side driver classes and the heap descriptor (`HBufC`), which is discussed further in Chapter 5.

This doesn't mean that the code is wrong – in many cases there are good reasons why they do not fit the theory. In the lower-level code, in particular, you'll find cases which may have been written before the name conventions were fully established or which, for efficiency reasons, have different characteristics and behavior. Whenever you come across a new class, it's worth comparing it to the rules above to see if it fits and, if not, considering why it doesn't. You'll be able to add a set of good exceptions to the rules to your list of cunning Symbian OS tricks, while discarding any code that unnecessarily contravenes the conventions – thus learning from others' mistakes.

1.8 Summary

This chapter reviewed the major class types used when coding for Symbian OS, and described their main features such as any special requirements when constructing or destroying objects of the class, whether they can be stack- or heap-based and typical member data contained by the class (if any). In particular, the chapter discussed the use of the class name conventions to indicate the cleanup characteristics of objects of the class in the event of a leave.

The guidelines within this chapter should be useful when writing a class – they can help you save time when coding and testing it, if only by cutting down on the rewrite time. If you can stick as closely as possible to Symbian OS conventions, your clients will know how you mean your class to be constructed, used and destroyed. This naturally benefits them, but can also help you reduce documentation and support-time further down the track.

2

Leaves: Symbian OS Exceptions

Go away. I'm all right

Said to be the last words of H. G. Wells

Symbian OS was first designed at a time when exceptions were not part of the C++ standard. Later, exception handling was introduced to the standard, but was found to add substantially to the size of compiled code and to run-time RAM overheads, regardless of whether or not exceptions were actually thrown. For these reasons, standard C++ exception handling was not considered suitable to add to Symbian OS, with its emphasis on a compact operating system and client code. When compiling Symbian OS code, the compilers are explicitly directed to disable C++ exception handling, and any use of the `try`, `catch` or `throw` keywords is flagged as an error.

An alternative to conventional, but rather awkward, error-checking around each function with a return value was needed. Thus "leaves"[1] were developed as a simple, effective and lightweight exception-handling mechanism which has become fundamental to Symbian OS. You'll encounter lots of "leaving code" when working on Symbian OS and you'll probably write some too. You need to know how to recognize code that leaves and how to use it efficiently and safely, since it's possible to leak memory inadvertently in the event of a leave. So when and how does a leave occur and why would you use it?

2.1 Leaving Functions

A leave may occur when you call a leaving function or if you explicitly call a system function to cause a leave. A leave is used to raise an

[1] "Leaves" is as in the verb "to leave" rather than the noun usually found attached to plants. A function that contains code which may leave is called a "leaving function" while code that has executed along the path of a leave (say, as the result of an exceptional condition) can be said to have "left".

exception and propagate an error value back up the call stack to a point at which it can be "caught" by a trap harness and handled appropriately. To all intents and purposes, code execution ends at the point of the leave and resumes where it is trapped. The leave sets the stack pointer to the context of a trap harness `TRAP` macro and jumps to the desired program location, restoring the register values. It does not terminate the flow of execution (unlike an assertion, which is used to detect programming errors and panic accordingly, as described in detail in Chapter 16).

`TRAP` and `User::Leave()` may be considered analogous to the standard library `setjmp()` and `longjmp()` methods respectively. A call to `setjmp()` stores information about the location to be "jumped to" in a jump buffer, which is used by `longjmp()` to determine the location to which the point of execution "jumps". A leave should only be used to propagate an exception to a point in the code which can handle it gracefully, unwinding the call stack as it does so. It should not be used to direct the normal flow of program logic.

A typical leaving function is one that performs an operation that is not guaranteed to succeed, such as allocation of memory, which may fail under low memory conditions. Since leaving functions by definition leave with an error code (a "leave code"), they do not also need to return error values. Indeed any error that occurs in a leaving function should be passed out as a leave; if the function does not leave it is deemed to have succeeded and will return normally. Generally, leaving functions should return `void` unless they use the return value for a pointer or reference to a resource allocated by the function. Later in this chapter, I'll discuss the factors that may influence your decision as to whether to implement a function that leaves or one that returns an error value.

Some examples of leaving function declarations are as follows:

```
void InitializeL();
static CTestClass* NewL();
RClangerHandle& CloneHandleL();
```

If a function may leave, its name must be suffixed with "L" to identify the fact. You must use this rule: of all Symbian OS naming conventions it is probably the most important. If you don't name a leaving function accordingly, callers of your code may not defend themselves against a leave and may potentially leak memory.

Functions may leave if they:

- call code that may leave without surrounding that call with a trap harness
- call one of the system functions that initiates a leave, such as `User::Leave()` or `User::LeaveIfError()`

- use the overloaded form of `operator new` which takes `ELeave` as a parameter (described in Section 2.2).

The suffix notation for names of functions which may leave is a simplification of the C++ exception specification which uses `throw(...)` or `throw(type)` by convention to indicate a function which may throw an exception. A call to `User::Leave()` or `User::LeaveIfError()` is similar to a C++ `throw` instruction (except for its destruction of stack-based variables, as I'll discuss shortly) while the `TRAP` macros are, in effect, a combination of `try` and `catch`.

`User::LeaveIfError()` tests an integer parameter passed into it and causes a leave (using the integer value as a leave code) if the value is less than zero, for example, one of the `KErrXXX` error constants defined in `e32std.h`. `User::LeaveIfError()` is useful for turning a non-leaving function which returns a standard Symbian OS error into one that leaves with that value.

`User::Leave()` doesn't carry out any value checking and simply leaves with the integer value passed into it as a leave code. `User::LeaveNoMemory()` also simply leaves but the leave code is hardcoded to be `KErrNoMemory` which makes it, in effect, the same as calling `User::Leave(KErrNoMemory)`.

`User::LeaveIfNull()` takes a pointer value and leaves with `KErrNoMemory` if it is NULL. It can sometimes be useful, for example, to enclose a call to a non-leaving function which allocates memory and returns a pointer to that memory or NULL if it is unsuccessful.

The following example shows four possible leaves:

```
TInt UseClanger(CClanger* aClanger);                  // Forward declaration

CClanger* InitializeClangerL()
   {
   CClanger* clanger = new (ELeave) CClanger();  // (1) Leaves if OOM
   CleanupStack::PushL(clanger);                 // (2) See Chapter 3
   clanger->InitializeL();                       // (3) May leave
   User::LeaveIfError(UseClanger(clanger));      // (4) Leaves on error
   CleanupStack::Pop(clanger);
   return (clanger);
   }
```

The L suffix is not checked during compilation so occasionally you may forget to append L to a function name, or may later add code to a previously non-leaving function which may then cause it to leave. Symbian OS provides a helpful tool, LeaveScan, that checks code for incorrectly-named leaving functions. It is described in more detail in Section 2.6.

If a function may leave, its name must be suffixed with "L".

2.2 Heap Allocation Using `new (ELeave)`

Let's take a closer look at the use of `new (ELeave)` to allocate an object on the heap. This overload leaves if the memory is unavailable, and thus allows the returned pointer to be used without a further test that the allocation was successful. We saw it used in the code fragment above:

```
CClanger* InitializeClangerL()
    {
    CClanger* clanger = new (ELeave) CClanger();
    CleanupStack::PushL(clanger);
    clanger->InitializeL();
    CleanupStack::Pop(clanger);
    return (clanger);
    }
```

The code above is preferable to the following code, which requires an additional check to verify that the `clanger` pointer has been initialized:

```
CClanger* InitializeClangerL()
    {
    CClanger* clanger = new CClanger();
    if (clanger)
        {
        CleanupStack::PushL(clanger);
        clanger->InitializeL();
        CleanupStack::Pop(clanger);
        }
    return (clanger);
    }
```

What exactly does `new (ELeave)` do? Well, when you call `new()` to allocate an object on the heap you are invoking the `new` operator. This first allocates the memory required for the object by calling `operator new` (yes, the naming scheme is incredibly confusing), passing in the size of the memory required. It then calls the constructor to initialize an object in that memory. This code is generated by the compiler, because it's not possible to call a constructor directly – which of course means that if you want an object constructed on the heap, you must allocate it, via the `new operator`, through a call to `new()`.

Symbian OS has overloaded the global `operator new` to take a `TLeave` parameter in addition to the size parameter provided implicitly by `new()`. The `TLeave` parameter is ignored by `operator new` and is only used to differentiate this form of `operator new` from the non-leaving version. The Symbian OS overload calls a heap allocation function that leaves if there is insufficient heap memory available:

```
// From e32std.h
enum TLeave {ELeave};
```

```
...
inline TAny* operator new(TUint aSize, TLeave);

// e32std.inl
inline TAny* operator new(TUint aSize, TLeave)
{return User::AllocL(aSize);}
```

Symbian OS has overloaded the global `operator new` to take a `TLeave` parameter. This overload leaves if memory is unavailable on the heap.

If a leaving function which allocates an object doesn't leave, the allocation was successful and there is no need to check the result further.

2.3 Constructors and Destructors

Before moving on to talk further about how to call leaving functions, let's consider which functions should *not* leave. Quite simply, neither a constructor nor a destructor should leave, since doing so would potentially leak memory and place the object upon which it is invoked in an indeterminate state.

Chapter 4 will discuss this in more detail, but essentially, if a constructor can fail, say, through lack of the resources necessary to create or initialize the object, you must remove the code that may leave from the constructor and use the two-phase construction idiom instead.

Likewise, a leave should not occur in a destructor or in cleanup code. One particular reason for this is that a destructor could itself be called as part of cleanup following a leave and a further leave at this point would be undesirable, if nothing else because it would mask the initial reason for the leave. More obviously, a leave part-way through a destructor will leave the object destruction incomplete which may leak its resources.

If a destructor must call a leaving function, it may potentially be trapped and the leave code discarded, although this causes problems when testing out of memory leaves using the debug macros described in Chapter 17. Generally, it's preferable to have a separate leaving function, which can be called before destruction, to perform actions that may fail and provide the caller an opportunity to deal with the problem before finally destroying the object. Typically these would be functions such as `CommitL()` or `FreeResourceL()`.

Constructors and destructors *must* not leave.

2.4 Working with Leaving Functions

Let's look at the practicalities of working with leaves. Below is an example of a call to a leaving function. You'll notice that there is no need to check that `ironChicken` is initialized before using it, since `CTestClass::NewL()` would have left if any failure had occurred.

```
void FunctionMayLeaveL()
    {
    // Allocates ironChicken on the heap
    CTestClass* ironChicken = CTestClass::NewL();
    // If NewL() didn't leave, ironChicken was allocated successfully
    ironChicken->FunctionDoesNotLeave();
    delete ironChicken;
    }
```

If the `CTestClass::NewL()` function leaves for some reason, it is the responsibility of that function to release any memory already allocated as part of the function. If successful, the function allocates, initializes and returns a heap-based object (`NewL()` functions and two-phase construction are discussed in more detail in Chapter 4). In the code above, a call to a non-leaving function follows, but consider the implications if a leaving function was called instead. For example:

```
void UnsafeFunctionL()
    {
    // Allocates test on the heap
    CTestClass* test = CTestClass::NewL();
    test->FunctionMayLeaveL(); // Unsafe – a potential memory leak!
    delete test;
    }
```

This is unsafe. Memory is allocated on the heap in the call to `CTest-Class::NewL()`, but the following function may leave. Should this occur `test` will not be deallocated; consequently the function has the potential to leak memory. In a scenario such as this, you should push the heap object onto the cleanup stack, which will delete it should a leave occur. The cleanup stack is described more fully in Chapter 3.

While heap variables referred to only by local variables may be orphaned in this way, member variables will not suffer a similar fate (unless their destructor neglects to delete them when it is called at some later point). Thus the following code is safe:

```
void CTestClass::SafeFunctionL()
    {
    iMember = CClangerClass::NewL(); // Allocates a heap member
    FunctionMayLeaveL();             // Safe
    }
```

Note that the `CTestClass` object (pointed to by "`this`" in `CTest-Class::SafeFunctionL()`) is not deleted in the event of a leave. The heap-based `iMember` is stored safely as a pointer member variable, to be deleted at a later stage with the rest of the object, through the class destructor.

I've shown that you must prevent leaks from the potential orphaning of heap-based local variables, but what about cleanup of stack variables if a leave occurs? The leave mechanism simply deallocates objects on the stack – it does not call any destructors they have defined as it does so, unlike a C++ `throw`. Stack objects that own a resource which must be deallocated, or otherwise "released" as part of destruction, would leak that resource in the event of a leave. Classes which are intended to be used on the stack must not need a destructor.

This is the reason why Symbian OS has a class naming convention which clearly defines the allowed usage of a class (described in Chapter 1). The only classes which may be instantiated and used safely on the stack are T classes, which the Symbian OS naming convention dictates must not have a destructor, and R classes, which do not have a destructor but use `Close()`, or a similar method, to free the associated resource. The cleanup stack must be used to ensure that this method is called in the event of a leave – I'll discuss how to do so in the next chapter.

```
class TMyClass
    {
public:
    TMyClass(TInt aValue);
private:
    TInt iValue;
    };

void AnotherSafeFunctionL()
    {
    TInt localInteger = 1; // Leave-safe (built-in type)
    FunctionMayLeaveL(localInteger);
    TMyClass localObject(localInteger); // Leave-safe object
    AnotherPotentialLeaverL(localObject);
    }
```

Let's consider what happens if you happen to have a local variable, an object of a T class, on the heap. In a leaving function, you still need to protect the heap memory from being orphaned by a leave but the object itself has no destructor. You'll recall that I mentioned earlier that the cleanup stack performs both destruction and deallocation upon the objects in its care in the event of a leave. Well, that's true for objects of class types which have destructors, but for T class objects, it simply deallocates the memory. There's more about this in Chapter 3.

```
void AnotherFunctionL()
    {
    TMyClass* localObject = new (ELeave) TMyClass();
    // Make localObject leave-safe using the cleanup stack
    CleanupStack::PushL(localObject);
    AnotherPotentialLeaverL(localObject);
    CleanupStack::PopAndDestroy(localObject);
    }
```

2.5 Trapping a Leave Using TRAP and TRAPD

Symbian OS provides two macros, TRAP and TRAPD, to trap a leave. The macros differ only in that TRAPD declares a variable in which the leave error code is returned, while the program code itself must declare a variable before calling TRAP. Thus the following statement:

```
TRAPD(result, MayLeaveL());
if (KErrNone!=result) // See footnote 2
    {
    // Handle error
    }
```

[2] Throughout this text, you'll notice that I prefer to use "back to front" comparisons in my if statements to prevent accidentally typing only a single =, which is valid C++ but isn't at all what is intended. Take the following example of the unforeseen consequences that can arise from this bug, which is often difficult to spot unless you have a helpful compiler that warns you about it.

```
TInt ContrivedFunction()
    { // The bug in this function means it will always return 0
    for (TInt index = 0; index < KContrivedValue; index++)
        {
        TInt calculatedValue = DoSomeComplexProcessing(index);
        // This assignment always returns true
        if (calculatedValue=KAnticipatedResult)
            return (index);
        }
    return (KErrNotFound);
    }
```

However, not everybody likes this style of coding. If you prefer not to use this technique, it pays to compile with a high warning level and pay attention to any resulting warnings.

is equivalent to:

```
TInt result;
TRAP(result, MayLeaveL());
if (KErrNone!=result)
    {
    // Handle error
    }
    ...
```

You should beware of nesting TRAPD macros and using the same variable name, as in the following snippet:

```
TRAPD(result, MayLeaveL())
if (KErrNone==result)
    {
    TRAPD(result, MayAlsoLeaveL())
    }
    ...
User::LeaveIfError(result);
```

In the example, two TInt result variables are declared, one for each TRAPD statement. The scope of the second result macro is bounded by the curly brackets that enclose it. Thus any leave code assigned to the second result variable, from the call to MayAlsoLeaveL(), is discarded on exiting the bounding brackets. The User::LeaveIfError() call thus only tests the leave code from the MayLeaveL() call, which is unlikely to be what the code intended. To ensure both values are tested, the second TRAPD should be replaced with a TRAP – thus reusing the TInt result declared by the initial TRAPD.

If a leave occurs inside the MayLeaveL() function, which is executed inside the harness, the program control will return immediately to the trap harness macro. The variable result will contain the error code associated with the leave (i.e. that passed as a parameter to the User::Leave() system function) or will be KErrNone if no leave occurred.

Any functions called by MayLeaveL() are executed within the trap harness, and so on recursively, and any leave that occurs during the execution of MayLeaveL() is trapped, returning the error code into result. Alternatively, TRAP macros can be nested to catch and handle leaves at different levels of the code, where they can best be dealt with. I'll discuss the runtime cost of using trap harnesses shortly, but if you find yourself using the TRAP macros several times in one function, or nesting a series of them, you may want to consider whether you can omit trapping all the leaving functions except at the top level, or change the layout of the code.

For example, there may be a good reason why the following function must not leave but needs to call a number of functions which may leave.

At first sight, it might seem straightforward enough simply to put each call in a trap harness.

```
TInt MyNonLeavingFunction()
    {
    TRAPD(result, FunctionMayLeaveL());
    if (KErrNone==result)
        TRAP(result,AnotherFunctionWhichMayLeaveL());
    if (KErrNone==result)
        TRAP(PotentialLeaverL());

    // Handle any error if necessary
    return (result);
    }
```

However, each `TRAP` has an impact in terms of executable size and execution speed. Both entry to and exit of a `TRAP` macro result in kernel executive calls[3] (`TTrap::Trap()` and `TTrap::UnTrap()`) being made. In addition, a `struct` is allocated at runtime to hold the current contents of the stack in order to return to that state if unwinding proves necessary in the event of a leave. Of course, use of the macro itself will also create additional inlined code, though it is probably insignificant in comparison. The combination of these factors does make a `TRAP` quite an expensive way of managing a leave. You should attempt to minimize the number of `TRAP`s you use in your code, where possible, either by allowing the leave to propagate to higher-level code or by making an adjustment similar to the following:

```
MyNonLeavingFunction()
    {
    TRAPD(result, MyLeavingFunctionL());
    // Handle any error if necessary
    return (result);
    }

void MyLeavingFunctionL()
    {
    FunctionMayLeaveL();
    AnotherFunctionWhichMayLeaveL();
    PotentialLeaverL();
    }
```

Of course, code is rarely as trivial as this example and you should beware of losing relevant error information. If a number of calls to potential leaving functions are packaged together as above and a leave

[3] A kernel executive call is made by user-side code to allow it to enter processor privileged mode in order to access kernel resources in a controlled manner. Control is switched to the kernel executive, and the processor is switched to supervisor mode, within the context of the calling thread.

code is returned from a call to the package of functions, it will not be clear which function left.

Every program (even a simple "hello world" application) must have at least one TRAP, if only at the topmost level, to catch any leaves that are not trapped elsewhere. If you are an application writer you don't need to worry about this, though, because the application framework provides a TRAP.

When testing any code that may leave, you should test both paths of execution, that is, for a successful call and for a call that leaves as a result of each of the exceptional conditions you expect to handle, for example, low memory conditions, failure to write to a file, etc. The Symbian OS macros to simulate such conditions are described in Chapter 17.

I mentioned earlier that, under normal circumstances, you shouldn't implement functions which both return an error and have the potential to leave. For example, consider a function, `OpenFileObjectL()`, that instantiates an object (`CFileObject`) which must be initialized with an open file handle. The implementation of this object may be in a separate library, the source code for which may not be available to the writer of this code.

```
TInt OpenFileObjectL(CFileObject*& aFileObject)
    {// File server session initialization is omitted for clarity
    ... // (handle is leave-safe)
    RFile file;
    TInt error = file.Open(...);
    if (KErrNone==error)
        {
        CleanupClosePushL(file); // Makes handle leave-safe
        // Propagates leaves from CFileObject
        aFileObject = CFileObject::NewL(file);
        CleanupStack::Pop(&file); // Owned by aFileObject now
        }
    return error; // Returns any error from RFile::Open()
    }
```

When a caller comes to use `OpenFileObjectL()`, they'll find that this function is more complex than it needs to be. At best, they may find themselves writing code something like the following:

```
void ClientFunctionL()
    {// Pass all errors up to be handled by a higher-level TRAP harness
    CFileObject* fileObject = NULL;
    TInt errorCode=OpenFileObjectL();
    if  (KErrNone!=errorCode)
        {
        User::Leave(errorCode);
        }
    ...
    }
```

Or they may use a TRAP to catch any leaves and return them as errors:

```
TInt ClientFunction()
    {
    CFileObject* fileObject = NULL;
    TInt errorCode;
    TRAPD(r, errorCode=OpenFileObjectL());
    if (KErrNone!=r)
        return (r);
    if (KErrNone!=errorCode)
        return (errorCode);
    ...
    }
```

Neither of these options is very attractive. Furthermore, if the client function can actually handle some of the errors at that point in the code, it becomes even messier. Should it differentiate between the two methods of propagating an error? And if so, how? Are the leaves in some way different to the errors returned? Could the same error sometimes be returned as an error value (say KErrNotFound from RFile::Open()) and sometimes as a leave code (say a leave from CFileObject::NewL() which attempts to read the contents of the file and finds it is missing some vital configuration data)?

In effect, this approach requires you to document the function clearly to allow callers to use it correctly and handle errors or exceptions as they choose. Maybe you can guarantee that you, or whoever takes responsibility for maintaining this function, keeps the documentation up to date. But you are returning an error directly from one, separate, component (class RFile) and a leave code from another (class CFileObject). Even if you know exactly what errors and leave codes each will use and are prepared to document them for callers of OpenFileObjectL(), you cannot guarantee that their error handling will not change.

It is preferable to restrict the code either to leaving or to returning an error. First, consider an implementation that leaves under all circumstances (notice that I've changed the signature to return the CFileObject rather than pass it as a reference-to-pointer parameter).

```
CFileObject* LeavingExampleL()
    {
    RFile file;
    User::LeaveIfError(file.Open(...));
    return (CFileObject::NewL(file));
    }
```

The implementation of the function is certainly smaller and less complex. If the calling code can handle some errors, then it can TRAP the method and switch on the error value returned in order to distinguish between the exceptional case and the non-exceptional case.

```
void ClientFunctionL()
   {
   CFileObject* fileObject = NULL;
   TRAPD(r, fileObject = LeavingExampleL());
   switch (r)
      {
      case (KErrNoMemory):
         ... // Free up some memory and try again...
      break;
         ...
      default:
         User::Leave(err);
      break;
      }
   }
```

Otherwise it can just call the function directly and allow a higher-level trap handler to handle all the leave codes:

```
CFileObject* fileObject = LeavingExampleL();
```

As an alternative to leaving on all errors, the function to instantiate a `CFileObject` could `TRAP` the call to `CFileObject::NewL()` and instead return an error for all failures:

```
TInt Example(CFileObject*& aFileObject)
   {
   RFile file;
   TInt error = file.Open(...);
   if (error == KErrNone)
      {
      TRAP(error, aFileObject = CThing::NewL(file));
      }
   return error;
   }
```

Again, if the calling code can handle some errors, it can switch on the return code; otherwise, it can call the function inside `User::LeaveIf-Error()` to pass all failures up to a `TRAP` harness in higher-level code.

```
CFileObject* fileObject = NULL;
User::LeaveIfError(Example(fileObject));
```

Which of the two implementations is preferable? Well, the leaving version is smaller and simpler and, as Martin Fowler discusses in *Refactoring: Improving the Design of Existing Code* (see Bibliography for further details), the use of exceptions clearly separates normal processing from error processing. Furthermore, the use of the alternative, error-returning function always incurs the overhead of a

TRAP, even if none of the callers can handle errors and they all call User::LeaveIfError() as shown in the previous code sample.

However, if you know who all your callers are (for example, if the function is internal to your component) and you know that the callers will all TRAP the call, it may be worthwhile implementing the version that returns an error. This limits the use of a TRAP to one place in the code.

Use of TRAP is an expensive way of managing a leave in terms of executable size and execution speed. You should attempt to minimize the number of TRAPs you use in your code where possible. However, every program must have at least one TRAP to catch any leaves that occur.

2.6 LeaveScan

LeaveScan is a useful tool which you should run regularly against your source code. It checks that all functions which have the potential to leave are named according to the Symbian OS convention, with a trailing L. LeaveScan can be used on your source to indicate areas of code where you may have forgotten to use the convention. By revealing where leaves may occur but are not acknowledged by the function name, it highlights potential bugs and gives you an opportunity to fix the problem and ensure that your code handles any leaves correctly.

LeaveScan works by examining each line of source code and checking that functions which do not terminate in L cannot leave. However, there are a few functions (more accurately, operators) in Symbian OS that may leave but cannot have an L suffix (such as operator<< and operator>> for RWriteStream and RReadStream respectively). The naming convention cannot be applied appropriately to operators and, unfortunately, LeaveScan does not have the sophisticated logic needed to recognize operators that may leave. When you use operators that you know have the potential to leave, you'll have to remember to check this code by sight yourself.

LeaveScan also checks functions which do have a trailing L to see if they really can leave. If functions are encountered which do not leave, LeaveScan raises a warning. However, this scenario can be perfectly valid, for example, when implementing an abstract function such as CActive::RunL(), some implementations may leave but others may not.

LeaveScan will highlight functions which may leave but are incorrectly named without a suffixed "L". The potential to leave occurs when a function:

- **calls other functions which themselves leave (and thus have function names ending in L) but does not surround the function call with a `TRAP` harness**
- **calls a system function which initiates a leave, such as `User::LeaveIfError()` or `User::Leave()`**
- **allocates an object on the heap using the Symbian OS overload of `operator new`.**

2.7 Summary

This chapter discussed leaves, which are the lightweight equivalent of C++ exceptions on Symbian OS. A leave is used to propagate an error which occurs because of exceptional conditions (such as being out of memory or disk space) to higher-level code which can handle it. A Symbian OS leave is equivalent to a C++ `throw` and a `TRAP` harness is used to catch the leave. In fact, a `TRAP` harness is effectively a combination of `try` and `catch`.

Having compared leaves and `TRAP`s with standard C++, it's worth making a comparison with the standard library too. `TRAP` and leave are analogous to the `setjmp()` and `longjmp()` methods, respectively – `setjmp()` stores information about the location to be "jumped to" in a jump buffer, which is used by `longjmp()` to direct the code to jump to that point.

On Symbian OS, if a function can leave, that is, fail under exceptional conditions, it indicates this by suffixing its function name with L. Of all the Symbian OS naming conventions, this is one you should comply with because, if you don't, it's hard to know whether you can call a function without potentially orphaning any local heap-based variables. You can check that you have adhered to the naming convention by running the LeaveScan tool over your source code.

A function can leave if it calls one of the system functions which cause a leave (such as `User::Leave()`), calls another leaving function (such as `NewL()`) or allocates memory using the Symbian OS leaving overload of `operator new`. Some functions should not leave, namely constructors (I'll discuss the reasons behind this in detail in Chapter 4, and explain how two-phase construction can be used to prevent it) and destructors, which could potentially leak memory by leaving before completing object cleanup.

This chapter described best practice for writing and calling leaving code, particularly when a function may have a return value for initialized pointer data. It also discussed the best use of the `TRAP` and `TRAPD` macros, which can have a significant overhead in terms of code size and runtime speed.

The next chapter discusses the use of the cleanup stack to protect heap-based local variables against orphaning in the event of a leave.

3

The Cleanup Stack

Life is pleasant. Death is peaceful. It's the transition that's troublesome
Jimi Hendrix

This chapter discusses a fundamental part of Symbian OS, the cleanup stack. Symbian OS is designed to perform well with limited memory, and part of that design must inevitably consider memory management when errors occur. The cleanup stack manages memory which would otherwise be "orphaned" (leaked) in the event of a leave.

But what, exactly, is "orphaning"? In the previous chapter, I described why Symbian OS doesn't use standard C++ exceptions, but instead handles exceptional conditions using "leaves". Code that can leave is, at some level, surrounded by a `TRAP`, or `TRAPD`, harness. If a leave occurs, control is transferred directly to the statement following the harness. In effect, the `TRAP` is equivalent to a `setjmp` and the leave to a `longjmp`. The stack memory is freed as the stack unwinds, but otherwise the stack frame is abandoned and no object destructors are called (which is unlike a standard C++ exception). This means that the destructors of any local variables or arguments passed by value, or objects created as member variables of either of these, will not be called. Some objects are "leave-safe" – they do not need destructors and contain only data which can be destroyed as stack memory is freed by the leave. This data may consist of the basic, built-in types or other objects which contain such types. In Symbian OS these are called T classes (T for "type", described in detail in Chapter 1), the characteristic of which is that they may safely be created on the stack in functions where code may leave.

If a local variable is a pointer to an object on the heap, when a leave occurs the pointer is destroyed without freeing that heap memory, which becomes unrecoverable, causing a memory leak. The memory is said to be orphaned. This means that C class objects, which are always created on the heap, as described in Chapter 1, are not leave-safe. Unless they are otherwise accessible for safe destruction (for example, as member variables of an object which is destroyed regardless of the leave), the memory they occupy on the heap and any resources they own are orphaned by a leave. R class objects are generally not leave-safe either, since the resources they own must be freed in the event of a leave (through a call to the appropriate `Close()` or `Release()` type function). If this call cannot made by an object accessible after the leave, the resource is orphaned.

Consider the following code, which creates an object of a C class (`CClanger`) on the heap, referenced only by an automatic variable, `clanger`. After creating the object, a function which may potentially leave, `InitializeL()`, is called. The heap-based object is not leave-safe and neither the heap memory it occupies, nor any objects it owns, are destroyed if `InitializeL()` leaves.[1]

```
void UnsafeFunctionL()
    {
    CClanger* clanger = new (ELeave) CClanger();
    clanger->InitializeL(); // Potential leaving function orphans clanger
    ... // Omitted for clarity
    delete clanger;
    }
```

You could remove the potential memory leak by placing a `TRAP`, or `TRAPD`, macro around the call to `InitializeL()` to catch any leaves. However, as Chapter 2 explained, the use of `TRAP`s should be limited, where possible, to optimize the size and run-time speed of the compiled binary. I have only shown one call which could cause the object to be leaked; in a typical function there may be other operations that may leave. It would be inefficient to surround each with a `TRAP`. From an error-handling perspective, too, it's often preferable to leave rather than return an error, as I discussed in Chapter 2.

When you combine both arguments, and consider the following irredeemably inefficient code, it's clear that there should be an alternative to the `TRAP` idiom to protect objects which are not inherently leave-safe:

```
void SafeButInefficientL()
    {
    CClanger* clanger = new (ELeave) CClanger();
    TRAPD(r,clanger->InitializeL()); // leave-safe, at a cost...
    if (KErrNone!=r)
```

[1] Before going any further, I should point out that the example is atypical of how a `CBase`-derived class should be instantiated. The Symbian OS coding standard recommends that all C classes wrap their construction and initialization code in a single static function, generally called `NewL()` or `NewLC()`, as illustrated later. To prevent a client from instantiating and initializing the object separately, as I've shown above, the standard advises that the constructors and initialization functions should be specified as protected in the class. This is known as two-phase construction, and is described further in Chapter 4.

```
        {
        delete clanger; // delete clanger to prevent a leak
        User::Leave(r); // leave with the same error... horrible!
        }

    // Now do something else that may leave
    TRAP(r, clanger->DoSomethingElseL());
    if (KErrNone!=r)
        {
        delete clanger; // Again, delete clanger to prevent a leak
        User::Leave(r);
        }

    ... // Omitted for clarity
    delete clanger;
    }
```

When a leave occurs, local variables are destroyed without first freeing any resources they own. The resource becomes unrecoverable, causing a memory leak, and is said to be "orphaned".

3.1 Using the Cleanup Stack

This is the cue for the cleanup stack, which is accessed through the static member functions of class `CleanupStack`, defined in `e32base.h`:

```
class CleanupStack
    {
public:
    IMPORT_C static void PushL(TAny* aPtr);
    IMPORT_C static void PushL(CBase* aPtr);
    IMPORT_C static void PushL(TCleanupItem anItem);
    IMPORT_C static void Pop();
    IMPORT_C static void Pop(TInt aCount);
    IMPORT_C static void PopAndDestroy();
    IMPORT_C static void PopAndDestroy(TInt aCount);
    IMPORT_C static void Check(TAny* aExpectedItem);
    inline static void Pop(TAny* aExpectedItem);
    inline static void Pop(TInt aCount, TAny* aLastExpectedItem);
    inline static void PopAndDestroy(TAny* aExpectedItem);
    inline static void PopAndDestroy(TInt aCount,
            TAny* aLastExpectedItem);
    };
```

Objects that are not otherwise leave-safe should be placed on the cleanup stack before calling code that may leave. This ensures they are destroyed correctly if a leave occurs; in the event of a leave, the cleanup stack manages the deallocation of all objects which have been placed upon it.

The following code illustrates a leave-safe but simpler version than the `SafeButInefficientL()` example:

```
void SafeFunctionL()
    {
    CClanger* clanger = new (ELeave) CClanger;

    // Push onto the cleanup stack before calling a leaving function
    CleanupStack::PushL(clanger);
    clanger->InitializeL();
    clanger->DoSomethingElseL()
    // Pop from cleanup stack
    CleanupStack::Pop(clanger);
    delete clanger;
    }
```

If `InitializeL()` succeeds, the lines of code following the call pop the automatic `CClanger` pointer from the cleanup stack and delete the object to which it points. This code could equally well be replaced by a single call to `CleanupStack::PopAndDestroy(clanger)` which performs both the pop and a call to the destructor in one step. The function is useful if the object is no longer required and may also be destroyed. And if `InitializeL()` or `DoSomethingElseL()` leave, the `clanger` object is destroyed by the cleanup stack itself as part of leave processing, which I'll discuss shortly.

Let's move on to consider some general good practices when working with the cleanup stack. Objects are pushed on and popped off the cleanup stack in strict order: a series of `Pop()` calls must occur in the reverse order of the `PushL()` calls. It is possible to `Pop()` or `PopAndDestroy()` one or more objects without naming them, but it's a good idea to name the object popping off. This averts any potential cleanup stack "imbalance" bugs which can occur when `Pop()` removes an object from the cleanup stack which was not the one intended.

```
void ContrivedExampleL()
    {// Note that each object is pushed onto the cleanup stack
    // immediately it is allocated, in case the succeeding allocation
    // leaves.

    CSiamese* sealPoint = NewL(ESeal);
    CleanupStack::PushL(sealPoint);
    CSiamese* chocolatePoint = NewL(EChocolate);
    CleanupStack::PushL(chocolatePoint);
    CSiamese* violetPoint = NewL(EViolet);
    CleanupStack::PushL(violetPoint);
    CSiamese* bluePoint = NewL(EBlue);
    CleanupStack::PushL(bluePoint);
```

```
sealPoint->CatchMouseL();
// Other leaving function calls, some of which use the cleanup stack
...

// Various ways to remove the objects from the stack and delete them:
// (1) All with one anonymous call - OK, but potentially risky
//     CleanupStack::PopAndDestroy(4);
// (2) All, naming the last object - Better
//     CleanupStack::PopAndDestroy(4, sealPoint);
// (3) Each object individually to verify the code logic
//     Note the reverse order of Pop() to PushL()
//     This is quite long-winded and probably unnecessary in this
//     example

CleanupStack::PopAndDestroy(bluePoint);
CleanupStack::PopAndDestroy(violetPoint);
CleanupStack::PopAndDestroy(chocolatePoint);
CleanupStack::PopAndDestroy(sealPoint);
}
```

At times, it's quite difficult to keep track of what's on the cleanup stack. Imbalance bugs can be difficult to find because they cause unexpected panics in code quite unrelated to where the error has occurred. To avoid them, it's best to name what you `Pop()` off the cleanup stack explicitly, if only in the early stages of coding until you're confident that the code is behaving as you expect. The checking is only performed in debug builds, so it has no impact on the speed of released code (although, if you do want to perform checking in a release build, you can use the `CleanupStack::Check()` function). If nothing else, it may help you track down bugs such as the one below where, under some conditions, an object is accidentally left on the cleanup stack when the function returns:

```
void AnotherContrivedExampleL(const TDesC& aString)
    {
    CClanger* clanger = AllocateClangerL(aString);
    CleanupStack::PushL(clanger);
    TBool result = NonLeavingFunction(clanger);
    if (!result)
        return;     // Whoops, clanger is still on the cleanup stack!

    DoSomethingThatLeavesL();
    CleanupStack::PopAndDestroy(clanger);
    }
```

But when should you `Pop()` an object from the cleanup stack? Well, it should never be possible for an object to be cleaned up more than once. If a pointer to an object on the cleanup stack is later stored elsewhere, say as a member variable of another object which is accessible after a leave, it should then be popped from the cleanup stack. If the pointer were retained on the cleanup stack, it would be destroyed by it, but the object storing the pointer to it would also attempt to destroy it, usually in its own destructor. Objects should be referred to either by another object or by the cleanup stack, but not by both.

```
void TransferOwnershipExampleL
    {
    CClanger* clanger = new (ELeave) CClanger();
    CleanupStack::PushL(clanger); // Pushed - the next function may leave
    iMemberObject->TakeOwnershipL(clanger);// iMemberObject owns it so
    CleanupStack::Pop(clanger);            // remove from cleanup stack
    }
```

Likewise, you should never push class member variables (objects prefixed by "i", if the Symbian OS naming convention is followed) onto the cleanup stack. The object may be accessed through the owning object which destroys it when appropriate, typically in its destructor. Of course, this requires the owning object to be leave-safe.

Incidentally, no panic occurs if you push or pop objects onto the cleanup stack more than once. The problem occurs if the cleanup stack tries to destroy the same object more than once, either through multiple calls to `PopAndDestroy()` or in the event of a leave. Multiple deletion of the same C class object will attempt to free memory which has already been released back to the heap, which causes a panic.

If an object is pushed onto the cleanup stack and remains on it when the function returns, the function name must have the suffix "C". This indicates to the caller that, when the function returns successfully, the cleanup stack has additional objects upon it. It is typically used by `CBase`-derived classes which define static functions to instantiate an instance of the class and leave it on the cleanup stack. The C suffix indicates to the caller of a function that it is not necessary to push any objects allocated by the function onto the cleanup stack. The following code creates an object of type `CSiamese` (which I used in an earlier example) and leaves it on the cleanup stack. This function is useful because the caller can instantiate `CSiamese` and immediately call a leaving function, without needing to push the allocated object back onto the cleanup stack:

```
CSiamese* CSiamese::NewLC(TPointColor aPointColour)
    {
    CSiamese* me = new (ELeave) CSiamese(aPointColour);
    CleanupStack::PushL(me); // Make this leave-safe...
    me->ConstructL();
    return (me) // Leave "me" on the cleanup stack for caller to Pop()
    }
```

However, a word of warning with regard to functions that leave objects on the cleanup stack – you must take care when calling them from inside a `TRAP` harness. If objects are pushed onto the cleanup stack inside a `TRAP` and a leave does not occur, they must be popped off again before exiting the macro, otherwise a panic occurs. This is because the cleanup stack stores nested levels of objects to destroy; each level is confined within a TRAP, and must be empty when the code inside it returns. Thus, the following code panics with E32USER-CBASE 71,[2] when it returns to the TRAPD macro.

```
CExample* MakeExample()
    {
    CExample* pExample = TRAPD(r, CExample::NewLC()); // This will panic
    return (pExample);
    }
```

Objects that are not leave-safe should be placed on the cleanup stack before calling code that may leave. The cleanup stack manages the deallocation of all objects which have been placed upon it in the event of a leave.

3.2 How Does the Cleanup Stack Work?

I'll use this section to go into how the cleanup stack works in some detail. If you came to this chapter only with an interest in how to use the cleanup stack, feel free to skip this section, since I have now covered the most important points for CBase-derived classes. Either read on anyway, or rejoin the chapter at Section 3.3, "Using the Cleanup Stack with non-CBase Classes", where I'll go into how to use the cleanup stack to make leave-safe objects of R classes, heap-based T classes and objects referenced through M class pointers.

As I've already described, the cleanup stack stores pointers to objects to be destroyed in the event of leave. The pointers are stored in nested "levels" associated with the TRAP macro under which they were pushed onto the stack. In the following code, if FeedL() leaves, only objects placed on the cleanup stack inside the TRAP macro are destroyed. Thus the CChilli object on the cleanup stack is destroyed, but the CExample object, stored on the cleanup stack prior to the call inside the TRAP macro, is not destroyed because the leave is only propagated to the level of the enclosing TRAP.

```
class CChilli;
class TCalibrateDragon
```

[2] The panic is described as follows in the panic documentation of the SDK: *"This panic is raised when TRAPs have been nested and an attempt is made to exit from a TRAP nest level before all the cleanup items belonging to that level have been popped off the cleanup stack."*

```
    {
public:
    FeedL(TChilliPepper aChilli);
private:
    void TasteChilliL(CChilli* aChilli);
    ... // Omitted for clarity
    };

CExample* CExample::NewL(TChilliPepper aChilli)
    {
    CExample* me = new (ELeave) CExample();
    CleanupStack::PushL(me); // Later functions may leave
    TCalibrateDragon dragon;
    TRAPD(r, dragon->FeedL(aChilli)); // TRAP this to act on the result

    // Do something dependent on whether FeedL() left
    if (KErrNone==r)
        {//... omitted for clarity
        }
    else
        {//... omitted for clarity
        }

    CleanupStack::Pop(me);
    return (me)
    }

void TCalibrateDragon::FeedL(TChilliPepper aChilli)
    {
    CChilli* chilli = CChilli::NewL(aChilli);
    CleanupStack::PushL(chilli);
    TasteChilliL(chilli); // Function leaves if chilli is too hot
    CleanupStack::PopAndDestroy(chilli);
    }
```

As the example shows, in the event of a leave, the cleanup stack destroys the objects it stores for that particular `TRAP` level. But how does it know how to destroy them? What does the cleanup stack class look like? How is the cleanup stack created? And what, the ever-cautious reader may ask, happens if `CleanupStack::PushL()` leaves?

Well, let's start at the beginning, the creation of a cleanup stack. You won't have to create a cleanup stack if you are writing a GUI application or a server, since both have cleanup stacks created for them as part of their framework code. However, if you are writing a simple console test application or creating a separate thread, and you need to use the cleanup stack or call code that does, you will need to allocate a cleanup stack, since if there is no cleanup stack allocated for a thread, a call to `CleanupStack::PushL()` causes an `E32USER-CBASE 32` panic.

```
CTrapCleanup* theCleanupStack = CTrapCleanup::New();
... // Code that uses the cleanup stack, within a TRAP macro
delete theCleanupStack;
```

The name given to the `CTrapCleanup` object is up to you, since the object is only referenced directly when you come to destroy it again. As you've already seen, access to the cleanup stack is through the static member functions of the `CleanupStack` class, defined in `e32base.h`. But how does this work?

Well, it's slightly confusing. When you create an object of type `CTrapCleanup`, the following occurs in `CTrapCleanup::New()`:

1. The current trap handler for the thread is stored.
2. Within `CTrapCleanup`, an `iHandler` object of type `TCleanupTrapHandler` is created (this owns a `CCleanup` object which actually contains the code which implements the cleanup stack).
3. A call to `User::SetTrapHandler()` then installs that `TCleanupTrapHandler` object as the new trap handler for the thread.

When you call `CleanupStack::PushL()` or `CleanupStack::Pop()`, the static functions call `User::TrapHandler()` to acquire the installed `TCleanupTrapHandler` object and gain access to the `CCleanup` object. Likewise, when a leave occurs, the trap handler catches it and invokes the `CCleanup` object to cleanup any objects it has stored (see Figure 3.1).

The `CleanupStack` class has three overloads of the `PushL()` function, which I'll discuss shortly. But first, let me allay any concerns you may have about the fact that `PushL()` can leave.

The reason that `PushL()` is a leaving function is because it may allocate memory to store the cleanup pointer and, thus, may fail in low memory situations. However, you don't need to worry that the object you are pushing onto the cleanup stack will be orphaned if a leave does occur. The cleanup stack is created with at least one spare slot. When you call `PushL()`, the pointer you pass is added to the vacant slot, and then, if there are no more vacant slots available, the cleanup stack code attempts to allocate one for the next `PushL()` operation. If that allocation fails, a leave occurs, but your pointer has already been stored safely and the object it refers to is safely cleaned up.

In fact, the cleanup stack code is more efficient than my simplistic explanation above; it allocates more than a single slot at a time. By default, it allocates four slots. In addition, it doesn't release slots when you `Pop()` objects out of them, so a `PushL()` frequently does not make any allocation and can be guaranteed to succeed. This can be useful in circumstances where you acquire more than one pointer that is not leave-safe to push onto the cleanup stack. You can use this knowledge to expand the cleanup stack to contain at least the number of slots you need, then create the objects and safely push them onto the vacant slots.

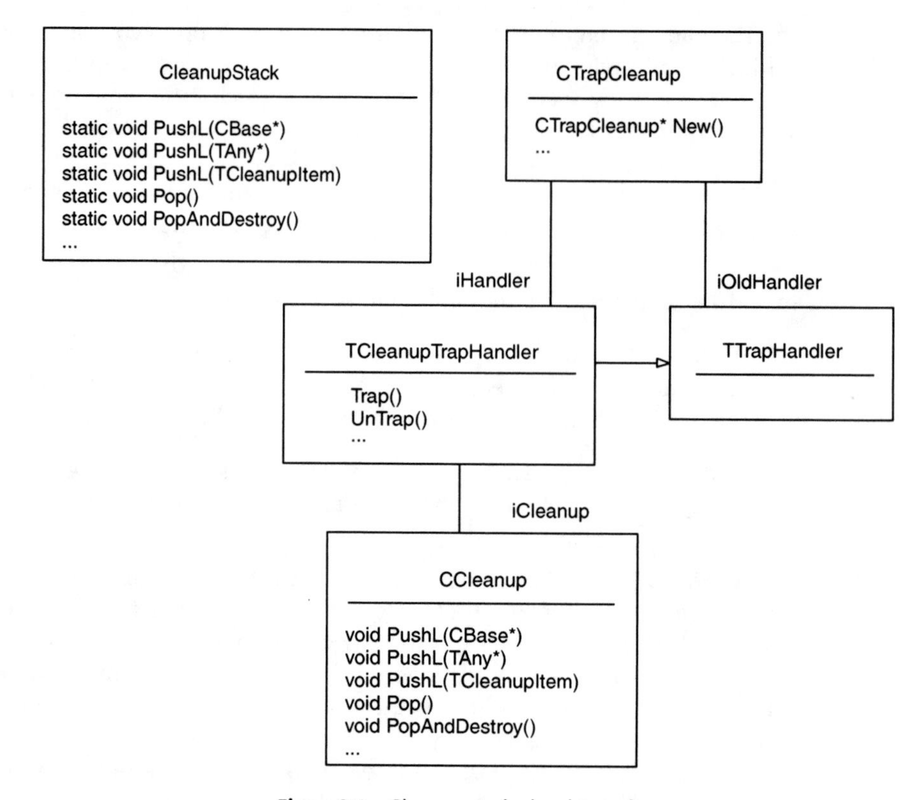

Figure 3.1 Cleanup stack class hierarchy

The cleanup stack stores pointers to objects to be destroyed using nested "levels" associated with the `TRAP` macro under which they were `PushL()`'d.

3.3 Using the Cleanup Stack with Non-`CBase` Classes

If you've rejoined after skipping the detailed section on how the cleanup stack works, welcome back. Up to this point, we've only really considered one of the overloads of the `CleanupStack::PushL()` function: `PushL(CBase* aPtr)`, which takes a pointer to a `CBase`-derived object. When `CleanupStack::PopAndDestroy()` is called for that object, or if leave processing occurs, the object is destroyed by invoking delete on the pointer, which calls the virtual destructor of the `CBase`-derived object. As mandated by C++, the object is destroyed by calling destructor code in the most-derived class first, moving up the inheritance hierarchy calling destructors in turn and eventually calling the empty CBase destructor. This is the reason that the CBase class has a virtual destructor – so that C class objects can be placed on the cleanup stack and destroyed safely if a leave occurs.

If you define a C class and forget to derive it from CBase, bad things happen if the cleanup stack tries to destroy it. If the object does not derive from CBase, the CleanupStack::PushL(TAny*) overload will be used to push it onto the cleanup stack, instead of the CleanupStack::PushL(CBase*) overload. Using this overload means that if the cleanup stack later comes to destroy the object, its heap memory is simply deallocated (by invoking User::Free()) and no destructors are called on it. In consequence, if your heap-based class does not inherit from CBase, directly or indirectly, and you push it onto the cleanup stack, its destructor is not called and it may not be cleaned up as you expect.

The CleanupStack::PushL(TAny*) overload is used when you push onto the cleanup stack a pointer to a heap-based object which does not have a destructor, for example, a T class object or a struct which has been allocated, for some reason, on the heap. Recall that T classes do not have destructors, and thus have no requirement for cleanup beyond deallocation of the heap memory they occupy.

Likewise, when you push heap descriptor objects (of class HBufC, described in Chapters 5 and 6) onto the cleanup stack, the CleanupStack::PushL(TAny*) overload is used. This is because HBufC objects are, in effect, heap-based objects of a T class. Objects of type HBufC require no destructor invocation, because they only contain plain built-in type data (a TUint representing the length of the descriptor and a character array of type TText8 or TText16). The only cleanup necessary is that required to free the heap cells for the descriptor object.

The third overload, CleanupStack::PushL(TCleanupItem), takes an object of type TCleanupItem, which allows you to put other types of object or those with customized cleanup routines onto the cleanup stack. A TCleanupItem object encapsulates a pointer to the object to be stored on the cleanup stack and a pointer to a function that provides cleanup for that object. The cleanup function can be a local function or a static method of a class.

For reference, here's the definition of the TCleanupItem class and the required signature of a cleanup function, which takes a single TAny* parameter and returns void:

```
// From e32base.h
typedef void (*TCleanupOperation)(TAny*);
//
class TCleanupItem
    {
public:
```

```
    inline TCleanupItem(TCleanupOperation anOperation);
    inline TCleanupItem(TCleanupOperation anOperation, TAny* aPtr);
private:
    TCleanupOperation iOperation;
    TAny* iPtr;
    friend class TCleanupStackItem;
    };
```

A call to `PopAndDestroy()` or leave processing removes the object from the cleanup stack and calls the cleanup function provided by the `TCleanupItem`.

```
class RSample
    {
public:
    static void Cleanup(TAny *aPtr);
    void Release();
    ... // Other member functions omitted for clarity
    };

// cleanup function
void RSample::Cleanup(TAny *aPtr)
    {// Casts the parameter to RSample and calls Release() on it
     // (error-checking omitted)
    (static_cast<RSample*>(aPtr))->Release();
    }

void LeavingFunctionL()
    {
    RSample theExample;
    TCleanupItem safeExample(RSample::Cleanup, &theExample);
    CleanupStack::PushL();

    // Some leaving operations called
    theExample.InitializeL(); // Allocates the resource
    ...

    // Remove from the cleanup stack and invoke RSample::Cleanup()
    CleanupStack::PopAndDestroy();
    }
```

You'll notice that the static `Cleanup()` function defined for `RSample` casts the object pointer it receives to a pointer of the class to be cleaned up. This is to allow access to the non-static `Release()` method which performs the cleanup. For code to be generic, the `TCleanupItem` object must store the cleanup pointer and pass it to the cleanup operation as a `TAny*` pointer.

The curious reader may be wondering how the cleanup stack discriminates between a `TCleanupItem` and objects pushed onto the cleanup stack using one of the other `PushL()` overloads. The answer is that it doesn't – all objects stored on the cleanup stack are of type `TCleanupItem`. The `PushL()` overloads that take `CBase` or `TAny` pointers construct a `TCleanupItem` implicitly and store it on the cleanup stack. As I've already described, the cleanup function for the `CBase*` overload deletes the `CBase`-derived object through its virtual destructor. For the `TAny*` overload, the cleanup function calls `User::Free()`, which simply frees the allocated memory.

Symbian OS also provides three utility functions, each of which generates an object of type `TCleanupItem` and pushes it onto the cleanup stack, for cleanup methods `Release()`, `Delete()` and `Close()`. The functions are templated so the cleanup methods do not have to be static.

CleanupReleasePushL(T& aRef)

Leave processing or a call to `PopAndDestroy()` will call `Release()` on an object of type T. To illustrate, here's the implementation from `e32base.inl`:

```
// Template class CleanupRelease
template <class T>
inline void CleanupRelease<T>::PushL(T& aRef)
    {CleanupStack::PushL(TCleanupItem(&Release,&aRef));}
template <class T>
void CleanupRelease<T>::Release(TAny *aPtr)
    {(STATIC_CAST(T*,aPtr))->Release();}
template <class T>
inline void CleanupReleasePushL(T& aRef)
    {CleanupRelease<T>::PushL(aRef);}
```

You may be wondering about the use of `STATIC_CAST` rather than the C++ standard `static_cast` I used in `RSample::Cleanup()`. Fear not, I'll explain why it's used at the end of this chapter, in Section 3.6 ("An Incidental Note on the Use of Casts").

And here's an example of its use to make an object referred to by an M class[3] pointer leave-safe. The M class does not have a destructor; instead the object must be cleaned up through a call to its `Release()` method, which allows an implementing class to customize cleanup:

```
class MExtraTerrestrial // See Chapter 1 for details of mixin classes
    {
public:
    virtual void CommunicateL() = 0;
    ... // The rest of the interface is omitted for clarity
    virtual void Release() = 0;
    };
```

---

[3] M classes are described in more detail in Chapter 1 and define a "mixin" interface which is implemented by a concrete inheriting class, which will, typically, also derive from `CBase`.

```
class CClanger : public CBase, MExtraTerrestrial
    {
public:
    static MExtraTerrestrial* NewL();
    virtual void CommunicateL(); // From MExtraTerrestrial
    virtual void Release();
private:
    CClanger();
    ~CClanger();
private:
    ...
    };

void TestMixinL()
    {
    MExtraTerrestrial* clanger = CClanger::NewL();
    CleanupReleasePushL(*clanger); // The pointer is deferenced here
    ... // Perform actions which may leave
    CleanupStack::PopAndDestroy(clanger); // Note pointer to the object
    }
```

Notice the asymmetry in the calls to push the object onto the cleanup stack and to pop it off again. As you can see from the definition above, `CleanupReleasePushL()` takes a reference to the object in question, so you must deference any pointer you pass it. However, if you choose to name the object when you `Pop()` or `PopAndDestroy()` it, you should pass a pointer to the object. This is because you're not calling an overload of `Pop()`.

The definition of `CleanupClosePushL()` is similarly asymmetric, but for `CleanupDeletePushL()` you should pass in a pointer. This is because objects which may be cleaned up using `Release()` or `Close()` may equally be `CBase`-derived heap-based objects, such as the one shown below, or R class objects which typically reside on the stack, as shown in a later example. As I'll describe, the `CleanupDeletePushL()` function should only apply to heap-based objects, so this function has a pointer argument.

CleanupDeletePushL(T& aRef)

Here's the implementation from `e32base.inl`:

```
// Template class CleanupDelete
template <class T>
inline void CleanupDelete<T>::PushL(T* aPtr)
    {CleanupStack::PushL(TCleanupItem(&Delete,aPtr));}
template <class T>
void CleanupDelete<T>::Delete(TAny *aPtr)
    {delete STATIC_CAST(T*,aPtr);}
template <class T>
inline void CleanupDeletePushL(T* aPtr)
    {CleanupDelete<T>::PushL(aPtr);}
```

As you can see, leave processing or a call to PopAndDestroy() calls delete on an object added to the cleanup stack using this method. A call is thus made to the destructor of the object and the heap memory associated with the object is freed. It is similar to using the CleanupStack::PushL() overload which takes a CBase pointer, and is useful when an M class-derived pointer must be placed on the cleanup stack.

Calling CleanupStack::PushL() and passing in an M class pointer to a heap-based object invokes the overload which takes a TAny* parameter. For this overload, the TCleanupItem constructed does not call the object's destructor, as I discussed above. Instead, it simply attempts to free the memory allocated for the object and, in doing so, causes a panic (USER 42) because the pointer does not point to the start of a valid heap cell.

This is because a concrete class that derives from an M class also inherits from another class, for example CBase, or a derived class thereof. The mixin pointer must be the second class in the inheritance declaration order if destruction of the object is to occur correctly, through the CBase virtual destructor. This means that when you access the object through the M class interface, the pointer has been cast to point at the M class sub-object which is some way into the allocated heap cell for the object. The memory management code cannot deduce this and panics when it cannot locate the appropriate structure it needs to free the cell back to the heap.

To resolve this, you should always use CleanupDeletePushL() to push M class pointers onto the cleanup stack, if the M class will be cleaned up through a virtual destructor. This creates a TCleanupItem for the object which calls delete on the pointer, using its destructor to cleanup the object correctly.

CleanupClosePushL(T& aRef)

If an object is pushed onto the cleanup stack using CleanupClosePushL(), leave processing or a call to PopAndDestroy() calls Close() on it. As an example, it can be used to make a stack-based handle to the file server leave-safe, closing it in the event of a leave:

```
void UseFilesystemL()
    {
    RFs theFs;
    User::LeaveIfError(theFs.Connect());
    CleanupClosePushL(theFs);
    ... // Call functions which may leave
    CleanupStack::PopAndDestroy(&theFs);
    }
```

`CleanupStack::PushL(TAny*)` is used to push a pointer to a heap-based object which does not have a destructor onto the cleanup stack (for example, a T class object or a struct).

`CleanupStack::PushL(TCleanupItem)` allows you to put other types of object, such as those with customized cleanup routines, onto the cleanup stack. Symbian OS provides three templated utility functions, each of which generates an object of type `TCleanupItem` and pushes it onto the cleanup stack, for `Release()`, `Delete()` and `Close()` cleanup methods.

3.4 Using `TCleanupItem` for Customized Cleanup

As I've already mentioned, `TCleanupItem` can be used to perform customized cleanup when a leave occurs. The cleanup function (`TCleanupOperation`) used by the `TCleanupItem` must be of the correct signature (`void Function(TAny* aParam)`) and may perform any cleanup action appropriate in the event of a leave. It is especially useful for transaction support where a sequence of functions must either succeed or fail as a single atomic operation. If any of the functions leaves, the object must be returned to its original state.

Alternatively, you can do this by putting a `TRAP` harness around each call that may leave, rolling back any previous changes when the `TRAP` "catches" a leave. Consider the following example code where an atomic method, `UpdateSpreadsheetL()`, calls two separate leaving functions inside `TRAP`s to recalculate the contents of a spreadsheet and update the view of the spreadsheet. If either of the methods leaves, the `TRAP`s catch the leave and allow the state of the spreadsheet and the spreadsheet view to be returned to the point before the `UpdateSpreadSheetL()` method was invoked, by calling `Rollback()`.

```
class CSpreadSheet : public CBase
    {
public:
    // Guarantees an atomic update
    void UpdateSpreadSheetL(const CUpdate& aUpdate);
    ... // Other methods omitted
private:
    // Recalculates the contents of iSpreadSheet
    void RecalculateL(const CData& aData);
    void UpdateL();    // Updates iSpreadSheetView
    void RollBack();   // Reverts to the previous state
private:
    CExampleSpreadSheet* iSpreadSheet;         // Defined elsewhere
    CExampleSpreadSheetView* iSpreadSheetView; // Defined elsewhere
```

```
   ...
   };

void CSpreadSheet::UpdateSpreadSheetL(const CUpdate& aUpdate)
   {
   TRAPD(result, RecalculateL(aData)); // Performs the recalculation
   if (KErrNone==result)              // if no error, update the view
      {
      TRAP(result, UpdateL());
      }
   if (KErrNone!=result) // RecalculateL() or UpdateL() failed
      {// Undo the changes
      RollBack();
      User::Leave(result);
      }
   }
```

However, as I've described, the use of TRAPs has an associated cost in terms of less efficient and more complex code. A preferable technique is to modify the code which reverts the object to its previous state. In the following example, the Rollback() method is now static and takes a TAny* pointer – and can be used to initialize a TCleanupItem. Rollback() casts the incoming TAny* to CSpreadSheet* and calls DoRollBack(), which returns the spreadsheet to its state before UpdateSpreadSheetL() was called.

```
class CSpreadSheet : public CBase
   {
public:
   // Guarantees an atomic update
   void UpdateSpreadSheetL(const CUpdate& aUpdate);
   ... // Other methods omitted
private:
   // Recalculates the contents of iSpreadSheet
   void RecalculateL(const CData& aData);
   void UpdateL(); // Updates iSpreadSheetView
      // Reverts to the previous state
   static void RollBack(TAny* aSpreadSheet);
   void DoRollBack();
 private:
   CExampleSpreadSheet* iSpreadSheet;
   CExampleSpreadSheetView* iSpreadSheetView;
   ...
   };

void CSpreadSheet::UpdateSpreadSheetL(const CUpdate& aUpdate)
   {
   // Initialize the TCleanupItem
   TCleanupItem cleanup(CSpreadSheet::RollBack, this);
   CleanupStack::PushL(cleanup);  // Put it on the cleanup stack
   RecalculateL(aData);
   UpdateL();
   CleanupStack::Pop(&cleanup);   // TCleanupItem is no longer required
   }
```

```
void CSpreadSheet::Rollback(TAny* aSpreadSheet)
    {
    ASSERT(aSpreadSheet);
    CSpreadSheet* spreadsheet = static_cast<CSpreadSheet*>(aSpreadSheet);
    spreadsheet->DoRollBack();
    }
```

3.5 Portability

The cleanup stack is a concept specific to Symbian OS, which means that code that uses it is non-portable. Sander van der Wal of mBrain Software has proposed using the cleanup stack to implement `auto_ptr<>`, part of the C++ standard, for Symbian OS, to make cleanup code more transferable. The `auto_ptr<>` template class can be used for automatic destruction, even when an exception is thrown and use of this standard cleanup mechanism could certainly allow code to be ported more rapidly. You can find more information about the implementation through links on the Symbian Developer Network (***www.symbian.com/developer***), where you can also find additional, in-depth articles about cleanup on Symbian OS.

3.6 An Incidental Note on the Use of Casts

C++ supports four cast operators (`static_cast`, `const_cast`, `reinterpret_cast` and `dynamic_cast`) which are preferable to the old C-style `cast` because they each have a specific purpose which the compiler can police for erroneous usage. For this reason, when you need to use a cast in Symbian OS code, you should endeavor to use `static_cast`, `const_cast` or `reinterpret_cast` rather than a basic C-style cast. Another benefit of using these casts is that they are more easily identified, both by the human eye and by search tools like grep.

Prior to Symbian OS v6.0, GCC did not support the C++ casting operators. Instead, Symbian OS defined a set of macros that simply used C-style casts for that compiler but at least allowed Symbian OS code to differentiate between the casts. The macros made the use of casts visible and allowed the code to be upgraded easily when the GCC compiler began to support the new casts, from v6.0 onwards. The macros were changed to reflect the availability of the new casts in v6.1.

You may encounter legacy code which still uses these macros, such as `CleanupReleasePushL()`, `CleanupDeletePushL()` and `CleanupClosePushL()`, but, from v6.0 onwards, you should simply use the native C++ casting operators rather than these macros.

You'll find the legacy macros defined in e32def.h:

```
#define CONST_CAST(type,exp) (const_cast<type>(exp))
#define STATIC_CAST(type,exp) (static_cast<type>(exp))
#define REINTERPRET_CAST(type,exp) (reinterpret_cast<type>(exp))
#if defined(__NO_MUTABLE_KEYWORD)
#define MUTABLE_CAST(type,exp) (const_cast<type>(exp))
#else
#define MUTABLE_CAST(type,exp) (exp)
#endif
```

You'll notice that dynamic_cast is not supported on Symbian OS. This cast is useful, allowing you to perform a safe cast from pointers of less derived classes to those in the inheritance hierarchy that are either more derived (child classes) or related across the inheritance tree. However, dynamic_cast makes use of runtime type identification (RTTI) which is not supported on Symbian OS because of the expensive runtime overhead required.[4] For this reason, dynamic_cast is, sadly, not supported by Symbian OS.

3.7 Summary

This chapter covered one of the most fundamental concepts of Symbian OS, the cleanup stack. It first explained the need for the cleanup stack, to prevent inadvertent heap memory leaks in the event of a leave. It then moved on to illustrate how, and how not, to use the cleanup stack most effectively. The chapter discussed the advantages of the cleanup stack over a TRAP harness, which can be used to prevent leaves but has a more significant impact on run-time speed and code size.

The chapter discussed in detail how the cleanup stack actually works and concluded by describing how it can best be used to ensure cleanup in the event of a leave for C class and R class objects, T class objects on the heap and objects referenced by M class interface pointers.

You can find more information about leaves in Chapter 2; the impact of leaves on constructor code is discussed in the next chapter.

[4] Another useful feature of dynamic_cast is that you can tell whether the cast succeeded; it will return a NULL pointer if you're casting a pointer or throw an exception if you're casting a reference. As you'll know from the previous chapter, Symbian OS does not support C++ exceptions, and uses leaves instead. This also contributes to the reason why dynamic_cast is unavailable on Symbian OS.

4

Two-Phase Construction

Either that wallpaper goes, or I do

Reputed to be the last words of Oscar Wilde

By now, you'll have gathered that Symbian OS takes memory efficiency very seriously indeed. It was designed to perform well on devices with limited memory, and uses memory management models such as the cleanup stack to ensure that memory is not leaked, even under error conditions or in exceptional circumstances.

Two-phase construction is an idiom that you'll see used extensively in Symbian OS code. To understand why it is necessary, we must consider exactly what happens when you create an object on the heap. Consider the following line of code, which allocates an object of type `CExample` on the heap, and sets the value of the `foo` pointer accordingly:

```
CExample* foo = new CExample();
```

The code calls the `new` operator, which allocates a `CExample` object on the heap if there is sufficient memory available. Having done so, it calls the constructor of class `CExample` to initialize the object.

As I described in Chapter 3, once you have a heap object such as that pointed to by `foo`, you can use the cleanup stack to ensure that the `CExample` object is correctly cleaned up in the event of a leave. But it is not possible to do so inside the `new` operator between allocation of the object and invocation of its constructor. If the `CExample` constructor itself leaves, the memory already allocated for the object and any memory the constructor may have allocated will be orphaned. This gives rise to a key rule of Symbian OS memory management: **no code within a C++ constructor should ever leave**.

Of course, to initialize an object you may need to write code that leaves, say to allocate memory or read a configuration file which may be missing or corrupt. It is under these circumstances that two-phase construction is useful.

When writing a new class you should break your construction code into two parts, or phases:

1. A basic constructor which cannot leave.
   It is this constructor which is called by the `new` operator. It implicitly calls base-class constructors and may also invoke functions that cannot leave and/or initialize member variables with default values or those supplied as arguments to the constructor.
2. A class method (typically called `ConstructL()`).
   This method may be called separately once the object, allocated and constructed by the `new` operator, has been pushed onto the cleanup stack; it will complete construction of the object and may safely perform operations that may leave. If a leave does occur, the cleanup stack calls the destructor to free any resources which have been successfully allocated and destroys the memory allocated for the object itself.

For example:

```
class CExample : public CBase
    {
public:
    CExample();          // Guaranteed not to leave
    ~CExample();         // Must cope with partially constructed objects
    void ConstructL();   // Second phase construction code - may leave
    ...
    };
```

The simplest implementation may expect clients to call the second-phase construction function themselves, but a class cannot rely on its objects being fully constructed unless it makes the call internally itself. A caller may forget to call the `ConstructL()` method after instantiating the object and, at the least, will find it a burden since it's not a standard C++ requirement. For this reason, it is preferable to make the call to the second phase construction function within the class itself. Obviously, the code cannot do this from within the simple constructor since this takes it back to the problem of having implemented a constructor which may leave.

A commonly used pattern in Symbian OS code is to provide a static function which wraps both phases of construction, providing a simple and easily identifiable means to instantiate objects of a class on the heap. The function is typically called `NewL()`, though a `NewLC()` function is often provided too, which is identical except that it leaves the constructed object on the cleanup stack for convenience.[1]

[1] In fact, `NewLC()` is such a commonly used idiom when working with Symbian OS that a very useful, independent Symbian developer website (***www.NewLC.com***) exists, providing news, tutorials and support forums.

```
class CExample : public CBase
    {
public:
    static CExample* NewL();
    static CExample* NewLC();
    ~CExample();         // Must cope with partially constructed objects
private:
    CExample();          // Guaranteed not to leave
    void ConstructL(); // Second phase construction code, may leave
    ...
    };
```

Note that the `NewL()` function is static, so you can call it without first having an existing instance of the class. The non-leaving constructors and second-phase `ConstructL()` functions have been made private[2] so a caller cannot instantiate objects of the class except through `NewL()`.

This prevents all of the following erroneous constructions:

```
CExample froglet; // BAD! C classes should not be created on the stack

CExample* myFroglet = new CExample(); // Caller must test for success
if (NULL!=myFroglet)
    {
    myFroglet->Hop(); // ConstructL() for myFroglet has not been called
    }

CExample* frogletPtr = new (ELeave) CExample();
frogletPtr->Hop(); // ConstructL() wasn't called, frogletPtr may not be
                   // fully constructed
```

Typical implementations of `NewL()` and `NewLC()` may be as follows:

```
CExample* CExample::NewLC()
    {
    CExample* me = new (ELeave) CExample(); // First phase construction
    CleanupStack::PushL(me);
    me->ConstructL(); // Second phase construction
    return (me);
    }

CExample* CExample::NewL()
    {
    CExample* me = CExample::NewLC();
    CleanupStack::Pop(me);
    return (me);
    }
```

---

[2] If you intend your class to be subclassed, you should make the default constructor protected rather than private so the compiler may construct the deriving classes. The `ConstructL()` method should be `private` (or `protected` if it is to be called by derived classes) to prevent clients of the class from mistakenly calling it on an object which has already been fully constructed.

Note that the `NewL()` function is implemented in terms of the `NewLC()` function rather than the other way around (which would be slightly less efficient since this would make an extra `PushL()` call on the cleanup stack).

Each function returns a fully constructed object, or will leave either if there is insufficient memory to allocate the object (that is, if the special Symbian OS overload of `operator new` leaves) or if the second phase `ConstructL()` function leaves. If second phase construction fails, the cleanup stack ensures both that the partially constructed object is destroyed and that the memory it occupies is returned to the heap.

The `NewL()` and `NewLC()` functions may, of course, take parameters with which to initialize the object. These may be passed to the simple constructor in the first phase or the second-phase `ConstructL()` function, or both.

If your class derives from a base class which also implements `ConstructL()`, you will need to ensure that this is also called, if necessary, when objects of your class are constructed (C++ will ensure that the simple first-phase constructors of your base classes are called). You should call the `ConstructL()` method of any base class explicitly (using the scope operator) in your own `ConstructL()` method, to ensure the base class object is fully constructed, before proceeding to initialize your derived object.

It is with class inheritance in mind that we can answer the following question: If it is possible to `PushL()` a partially constructed object onto the cleanup stack in a `NewL()` function, why not do so at the beginning of a standard constructor (thus allowing it to leave), calling `Pop()` when construction is complete? At first sight, this may be tempting, since the single-phase construction I described as unsafe at the beginning of the chapter would then be leave-safe, as long as the object was pushed onto the cleanup stack before any leaving operations were called in the constructor. However, if the class is to be used as a base class, the constructor of any class derived from it will incur one `PushL()` (and corresponding `Pop()`) in the constructor called at each level in the inheritance hierarchy, rather than a single cleanup stack operation in the `NewL()` function. In addition, from a cosmetic point of view, a C++ constructor cannot be marked with a suffixed `L` to indicate its potential to leave unless the class is itself named as such.

Before closing this chapter, it is worth noting that, when implementing the standard Symbian OS two-phase construction idiom, you should consider the destructor code carefully. Remember that a destructor must be coded to release all the resources that an object owns. However, the destructor may be called to cleanup partially constructed objects if a leave occurs in the second-phase `ConstructL()` function. The destructor code cannot assume that the object is fully initialized and you should beware of calling functions on pointers which may not yet be set to point to valid objects. Of course, the memory for a CBase-derived object is guaranteed to be set to binary zeroes on first construction (as described in Chapter 1). It is safe for a destructor to call delete on a NULL pointer, but you should beware of attempting to free other resources without checking whether the handle or pointer which refers to them is valid, for example:

```
CExample::~CExample()
   {
   if (iMyAllocatedMember)
      {
      iMyAllocatedMember->DoSomeCleanupPreDestruction();
      delete iMyAllocatedMember;
      }
   }
```

On Symbian OS, a C++ constructor should never leave, since any memory allocated for the object (and any memory the constructor may already have allocated) would be orphaned by the leave. Instead, construction code should be broken into two phases within a public static member function, typically called NewL() or NewLC().

4.1 Summary

A constructor should never be able to leave, because if it is called and leaves when an object is instantiated on the heap that object will be orphaned, causing a memory leak. For this reason, two-phase construction is used extensively in Symbian OS code for CBase-derived classes. It provides a means by which heap-based objects may be instantiated without the need to prevent construction and initialization code from leaving.

The first phase of two-phase construction allocates an object on the heap and may perform basic initialization which cannot leave. The second phase pushes the object onto the cleanup stack, to ensure it will be cleaned up in the event of a leave, before calling any further construction code which may leave.

Two-phase construction is generally performed by methods internal to the class rather than exposed to a caller, who may not appreciate that both phases of construction are required. Typically, two-phase construction is performed on Symbian OS using static NewL() and NewLC() methods (the latter leaves the constructed object on the cleanup stack). The second-phase construction method is usually called ConstructL() or InitializeL() and is usually specified as `protected` or `private` – as are the constructors – which enforces two-phase construction as the only means by which an object can be instantiated.

Two-phase construction is typically used for C classes, since T classes do not usually require complex construction code (because they do not contain heap-based member data) and R classes are usually created uninitialized, requiring their callers to call `Connect()` or `Open()` to associate the R class object with a particular resource. You can find more information about the characteristics of the various Symbian OS class types in Chapter 1, which discusses them in detail.

5

Descriptors: Symbian OS Strings

Get your facts first, then you can distort them as you please

Mark Twain

The Symbian OS string is known as a "descriptor", because it is self-describing. A descriptor holds the length of the string of data it represents as well as its "type", which identifies the underlying memory layout of the descriptor data. Descriptors have something of a reputation among Symbian OS programmers because they take some time to get used to. The key point to remember is that they were designed to be very efficient on low memory devices, using the minimum amount of memory necessary to store the string, while describing it fully in terms of its length and layout. There is, necessarily, some trade-off between efficiency and simplicity of use, which this chapter illustrates. The chapter is intended to give a good understanding of the design and philosophy of descriptors. The next chapter will show how to use descriptors most effectively by looking at some of the more frequently used descriptor API functions and describing some common descriptor mistakes and misconceptions.

Descriptors have been part of Symbian OS since its initial release and they have a well established base of documentation. Despite this, they can still appear confusing at first sight, perhaps because there are quite a number of descriptor classes, all apparently different although interoperable.[1] They're not like standard C++ strings, Java strings or the MFC `CString` (to take just three examples) because their underlying memory allocation and cleanup must be managed by the programmer. But they are not like C strings either; they protect against buffer overrun and don't rely on NULL terminators to determine the length of the string. So let's discuss what they are and how they work – initially by looking at a few concepts before moving on to the different descriptor classes.

First, I should make the distinction between descriptors and literals; the latter can be built into program binaries in ROM because they

[1] To paraphrase Andrew Tanenbaum: The nice thing about descriptors is that there are so many to choose from.

are constant. Literals are treated a bit differently to descriptors and I'll come back to them later in the chapter. For now, the focus is on descriptors.

Another issue is the "width" of the string data, that is, whether an individual character is 8 or 16 bits wide. Early releases, up to and including Symbian OS v5, were narrow builds with 8-bit native characters, but since that release Symbian OS has been built with wide 16-bit characters as standard, to support Unicode character sets. The operating system was designed to manage both character widths from the outset by defining duplicate sets of descriptor classes for 8- and 16-bit data. The behavior of the 8- and 16-bit descriptor classes is identical except for `Copy()` and `Size()`, both of which are described in the next chapter. In addition, a set of neutral classes are `typedef`'d to either the narrow or wide descriptor classes, depending on the build width. You can identify the width of a class from its name. If it ends in 8 (e.g. `TPtr8`) it assumes narrow 8-bit characters, while descriptor class names ending with 16 (e.g. `TPtr16`) manipulate 16-bit character strings. The neutral classes have no number in their name (e.g. `TPtr`) and, on releases of Symbian OS since v5u,[2] they are implicitly wide 16-bit strings.

The neutral classes were defined for source compatibility purposes to ease the switch between narrow and wide builds. Although today Symbian OS is always built with 16-bit wide characters, you are well advised to continue to use the neutral descriptor classes where you do not need to state the character width explicitly.

Descriptors can also be used for binary data because they don't rely on a NULL terminating character to determine their length. The unification of binary and string-handling APIs makes it easier for programmers and, of course, the ability to re-use string manipulation code on data helps keep Symbian OS compact. To work with binary data, you need to code specifically with the 8-bit descriptor classes. The next chapter discusses how to manipulate binary data in descriptors in more detail.

So, with that knowledge in hand, we can move on to consider the descriptor classes in general.

5.1 Non-Modifiable Descriptors

All (non-literal) descriptors derive from the base class `TDesC` which is `typedef`'d to `TDesC16` in `e32std.h` and defined in `e32des16.h` (the narrow version, `TDesC8`, can be found in `e32des8.h`). Chapter 1 discusses Symbian OS class naming conventions and explains what the "T" prefix represents. The "C" at the end of the class name is more

[2] Symbian OS v5u was used in the Ericsson R380 mobile phone. This version is also sometimes known as "ER5U", which is an abbreviation of "EPOC Release 5 Unicode".

relevant to this discussion, however; it reflects that the class defines a **non-modifiable** type of descriptor, whose contents are **constant**. The class provides methods for determining the length of the descriptor and accessing the data.

The length of the descriptor is returned, unsurprisingly, by the `Length()` method. The layout of every descriptor object is the same, with 4 bytes holding the length of the data it currently contains. (Actually, only 28 of the available 32 bits are used to hold the length of the descriptor data; 4 bits are reserved for another purpose, as I'll describe very shortly. This means that the maximum length of a descriptor is limited to 2^{28} bytes, 256 MB, which should be more than sufficient!)

The `Length()` method in `TDesC` is never overridden by its subclasses since it is equally valid for all types of descriptor. However, access to the descriptor data is different depending on the implementation of the derived descriptor classes but Symbian OS does not require each subclass to implement its own data access method using virtual functions. It does not use virtual function overriding because this would place the burden of an extra 4 bytes on each derived descriptor object, added by C++ as a virtual pointer (`vptr`) to access the virtual function table. As I've already described, descriptors were designed to be as efficient as possible and the size overhead to accommodate a `vptr` was considered undesirable. Instead, to allow for the specialization of derived classes, the top 4 bits of the 4 bytes that store the length of the descriptor object are reserved to indicate the type of descriptor.

There are currently five derived descriptor classes, each of which sets the identifying bits as appropriate upon construction. The use of 4 bits to identify the type limits the number of different types of descriptor to 2^4 (=16), but since only five types have been necessary in current and previous releases of Symbian OS, it seems unlikely that the range will need to be extended significantly in the future.

Access to the descriptor data for all descriptors goes through the non-virtual `Ptr()` method of the base class, `TDesC`, which uses a `switch` statement to check the 4 bits, identify the type of descriptor and return the correct address for the beginning of its data. Of course, this requires that the `TDesC` base class has knowledge of the memory layout of its subclasses hardcoded into `Ptr()`.

With the `Length()` and `Ptr()` methods, the `TDesC` base class can implement all the operations you'd typically expect to perform on a constant string (such as data access, comparison and search). Some of these methods are described in detail in the next chapter, and all will be documented in full in your preferred SDK. The derived classes all inherit these methods and, in consequence, all constant descriptor manipulation is performed by the same base class code, regardless of the type of the descriptor.

The non-modifiable descriptor class `TDesC` is the base class from which all non-literal descriptors derive. It provides methods to determine the length of the descriptor and to access its data. In addition, it implements all the operations you'd typically expect to perform on a constant string.

5.2 Modifiable Descriptors

Let's now go on to consider the **modifiable** descriptor types, which all derive from the base class `TDes`, itself a subclass of `TDesC`. `TDes` has an additional member variable to store the maximum length of data allowed for the current memory allocated to the descriptor. The `MaxLength()` method of `TDes` returns this value. Like the `Length()` method of `TDesC`, it is not overridden by the derived classes.

`TDes` defines the range of methods you'd expect for modifiable string data, including those to append, fill and format the descriptor data. Again, all the manipulation code is inherited by the derived classes, and acts on them regardless of their type. Typically, the derived descriptors only implement specific methods for construction and copy assignment.

None of the methods allocates memory, so if they extend the length of the data in the descriptor, as `Append()` does, for example, you must ensure that there is sufficient memory available for them to succeed before calling them. Of course, the length of the descriptor can be less than the maximum length allowed and the contents of the descriptor can shrink and expand, as long as the length does not exceed the maximum length. When the length of the descriptor contents is shorter than the maximum length, the final portion of the descriptor is simply unused.

The modification methods use assertion statements to check that the maximum length of the descriptor is sufficient for the operation to succeed. These will panic if an overflow would occur if they proceeded, allowing you to detect and fix the programming error (panics are described in detail in Chapter 15 and assertions in Chapter 16).

The very fact that you can't overflow the descriptor makes the code robust and less prone to hard-to-trace memory scribbles. In general, descriptor classes use `__ASSERT_ALWAYS` to check that there is sufficient memory allocated for an operation, raising a `USER` category panic if the assertion fails. In the event of such a panic, it can be assumed that no illegal access of memory has taken place and that no data was moved or corrupted.

The base classes provide and implement the APIs for constant and modifiable descriptor operations for consistency, regardless of the actual type of the derived descriptor. For this reason, the base classes should be used as arguments to functions and return types, allowing descriptors to be passed around in code without forcing a dependency on a particular type. However, if you attempt to create objects of type `TDesC` and `TDes` you'll find that they cannot be instantiated directly because their default constructors are protected.[3]

So it's to the derived descriptor types that we now turn, since these are the descriptor classes that you'll actually instantiate and use. As I mentioned earlier, it can at first sight appear quite confusing because there is a proliferation of descriptor classes. I've already explained why there are three versions of each class, e.g. `TDes8`, `TDes16` and `TDes`, for narrow, wide and neutral (implicitly wide) classes respectively. Let's now look at the main descriptor types, initially considering their general layout in memory before moving on to look at each class in more detail. I'll describe the differences between the classes and the methods each defines over and above those provided by the `TDesC` and `TDes` base classes. The following chapter will go further into how to use the base class APIs, as well as noting any useful tips or mistakes commonly made when doing so. For comprehensive information about the descriptor APIs, you should refer to the SDK documentation.

As I'll describe, descriptors come in two basic layouts: pointer descriptors, in which the descriptor holds a pointer to the location of a character string stored elsewhere, and buffer descriptors, where the string of characters forms part of the descriptor.

`TDes` is the base class for all modifiable descriptors, and itself derives from `TDesC`. It has a method to return the maximum amount of memory currently allocated to hold data, and a range of methods for modifying string data.

When using descriptors, memory management is your responsibility. Descriptors do not perform allocation, re-allocation or garbage collection, because of the extra overhead that would carry. However, descriptor functions do check against access beyond the end of the data, and will panic if passed out-of-bounds parameters.

---

[3] There is no copy constructor declared, so the compiler generates a default public version which can be used to instantiate a `TDes` or `TDesC` by copy, although you are unlikely to have a valid reason for doing this:

```
_LIT(KExampleLiteral, "The quick brown fox jumps over the lazy dog");
TPtrC original(KExampleLiteral);
TDesC copy(original); // Shallow copy the type, length & data
```

5.3 Pointer Descriptors

The string data of a pointer descriptor is separate from the descriptor object itself and can be stored in ROM, on the heap or on the stack. The memory that holds the data is not "owned" by the descriptor and is not managed through it. Thus, if it is on the heap, the memory is created, reallocated if necessary, and destroyed using a heap descriptor pointer (`HBufC`, described below). If a pointer descriptor is referencing a stack-based string, the memory in question will already have been allocated on the stack. The pointer descriptors themselves are usually stack-based, but they can be used on the heap, for example as a member variable of a `CBase`-derived class. Pointer descriptors are agnostic about where the memory they point to is actually stored.

In a non-modifiable pointer descriptor (`TPtrC`), the pointer to the data follows the length word, thus the total size of the descriptor object is two words. In a modifiable pointer descriptor (`TPtr`), it follows the maximum length word and the descriptor object is three words in length. Figure 5.1 compares the memory layouts of `TPtr` and `TPtrC`.

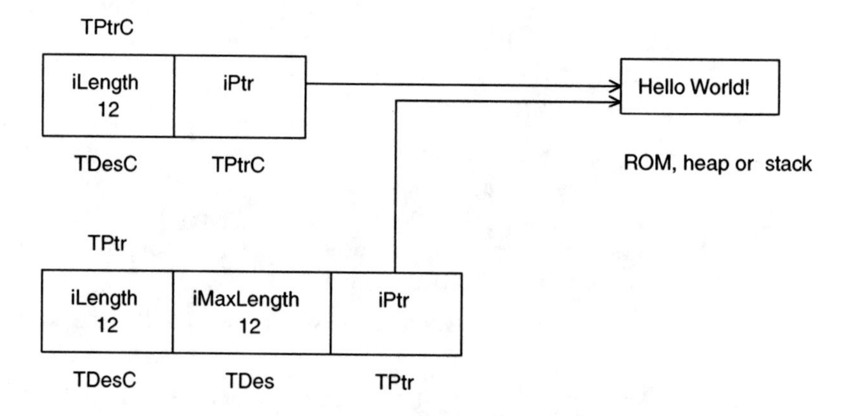

Figure 5.1 Memory layouts of pointer descriptors

TPtrC

`TPtrC` is the equivalent of using `const char*` when handling strings in C. The data can be accessed but not modified: that is, the data in the descriptor is constant. All the non-modifiable operations defined in the `TDesC` base class are accessible to objects of type `TPtrC`. The class also defines a range of constructors to allow `TPtrC` to be constructed from another descriptor, a pointer into memory or a zero-terminated C string.

```
// Literal descriptors are described later in this chapter
_LIT(KLiteralDes, "Sixty zippers were quickly picked from the woven
      jute bag");

TPtrC pangramPtr(KLiteralDes); // Constructed from a literal descriptor
TPtrC copyPtr(pangramPtr);     // Copy constructed from another TPtrC

TBufC<100> constBuffer(KLiteralDes); // Constant buffer descriptor
TPtrC ptr(constBuffer);              // Constructed from a TBufC

// TText8 is a single (8-bit) character, equivalent to unsigned char
const TText8* cString = (TText8*)"Waltz, bad nymph, for quick jigs
      vex";
// Constructed from a zero-terminated C string
TPtrC8 anotherPtr(cString);

TUint8* memoryLocation; // Pointer into memory initialized elsewhere
TInt length;            // Length of memory to be represented
...
TPtrC8 memPtr(memoryLocation,length); // Constructed from a pointer
```

The pointer itself may be changed to point at different string data – the `Set()` methods in `TPtrC` are defined for that purpose. If you want to indicate that the data your `TPtrC` points at should not be changed, you can declare the `TPtrC` to be `const`, which typically generates a compiler warning if an attempt is made to call `Set()` upon it. It will not fail, however, since the rules of `const`-ness in C++ are such that both `const` and non-`const` functions may be called on a `const` object.

```
// Literal descriptors are described later in this chapter
_LIT(KLiteralDes1, "Sixty zippers were quickly picked from the woven jute
      bag");
_LIT(KLiteralDes2, "Waltz, bad nymph, for quick jigs vex");

TPtrC alpha(KLiteralDes1);
TPtrC beta(KLiteralDes2);
alpha.Set(KLiteralDes2); // alpha points to the data in KLiteralDes2
beta.Set(KLiteralDes1);  // beta points to the data in KLiteralDes1

const TPtrC gamma(beta); // Points to the data in beta, KLiteralDes1
gamma.Set(alpha);        // Generates a warning, but points to alpha
```

TPtr

`TPtr` is the modifiable pointer descriptor class for access to and modification of a character string or binary data. All the modifiable and non-modifiable base class operations of `TDes` and `TDesC` respectively may be performed on a `TPtr`.

The class defines constructors to allow objects of type `TPtr` to be constructed from a pointer into an address in memory, setting the length and maximum length as appropriate.

The compiler also generates implicit default and copy constructors, since they are not explicitly declared protected or private in the class. A `TPtr` object may be copy constructed from another modifiable pointer descriptor, for example, by calling the `Des()` method on a non-modifiable buffer, which returns a `TPtr` as shown below:

```
_LIT(KLiteralDes1, "Jackdaws love my big sphinx of quartz");
TBufC<60> buf(KLiteralDes1);  // TBufC are described later


TPtr ptr(buf.Des()); // Copy construction; can modify the data in buf
TInt length = ptr.Length();            // Length = 12
TInt maxLength = ptr.MaxLength(); // Maximum length = 60, as for buf

TUint8* memoryLocation;                // Valid pointer into memory
...
TInt len = 12;                         // Length of data to be represented
TInt maxLen = 32;                      // Maximum length to be represented

// Construct a pointer descriptor from a pointer into memory
TPtr8 memPtr(memoryLocation, maxLen); // length = 0, max length = 32
TPtr8 memPtr2(memoryLocation, len, maxLen); // length = 12, max = 32
```

In addition, the class provides an assignment operator, `operator=()`, to copy data into the memory referenced by the pointer (from another modifiable pointer descriptor, a non-modifiable pointer or a zero-terminated string). If the length of the data to be copied exceeds the maximum length of the descriptor, a panic will be raised. Like `TPtrC`, this class also defines a `Set()` method to change the descriptor to point at different data.

```
_LIT(KLiteralDes1, "Jackdaws love my big sphinx of quartz");
TBufC<60> buf(KLiteralDes1);  // TBufC are described later
TPtr ptr(buf.Des());          // Points to the contents of buf

TUint16* memoryLocation;      // Valid pointer into memory
...
TInt maxLen = 40; // Maximum length to be represented
TPtr memPtr(memoryLocation, maxLen); // length = 12, max length = 40

// Copy and replace
memPtr = ptr; // memPtr data is KLiteralDes1 (37 bytes), maxLength = 40

_LIT(KLiteralDes2, "The quick brown fox jumps over the lazy dog");
TBufC<100> buf2(KLiteralDes2);  // TBufC are described later
TPtr ptr2(buf2.Des());          // Points to the data in buf
```

```
// Replace what ptr points to
ptr.Set(ptr2); // ptr points to contents of buf2, max length = 100
memPtr = ptr2; // Attempt to update memPtr which panics because the
// contents of ptr2 (43 bytes) exceeds max length of memPtr (40 bytes)
```

You should be careful not to confuse `Set()`, which resets your descriptor to point at a new data area (with corresponding modification to the length and maximum length members) with `operator =()` which merely copies data into the existing descriptor (and may modify the descriptor length but not its maximum length).

5.4 Stack-Based Buffer Descriptors

The stack-based buffer descriptors may be modifiable or non-modifiable. The string data forms part of the descriptor object, located after the length word in a non-modifiable descriptor and after the maximum length word in a modifiable buffer descriptor. Figure 5.2 compares the memory layouts of `TBuf` and `TBufC`.

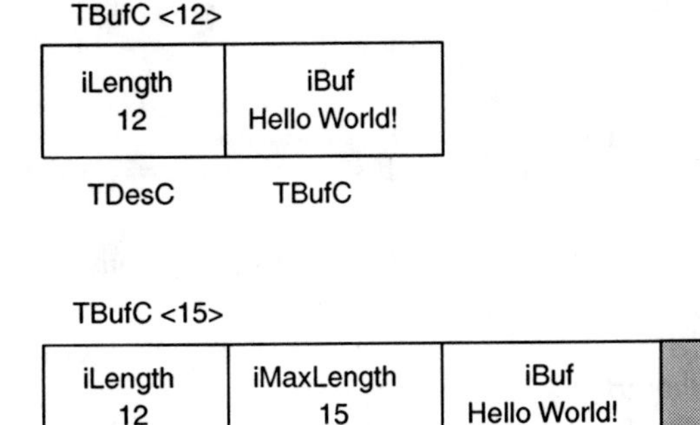

Figure 5.2 Buffer descriptors

These descriptors are useful for fixed-size or relatively small strings, say up to the length of a 256-character filename. Being stack-based, they should be used when they have a lifetime that coincides with that of their creator. They may be considered equivalent to `char[]` in C, but with the benefit of overflow checking.

TBufC<n>

This is the non-modifiable buffer class, used to hold constant string or binary data. The class derives from TBufCBase (which derives from TDesC, and exists as an inheritance convenience rather than to be used directly). TBufC<n> is a thin template class which uses an integer value to determine the size of the data area allocated for the buffer descriptor object. Chapter 19 describes the thin template pattern and its role in Symbian OS code.

The class defines several constructors that allow non-modifiable buffers to be constructed from a copy of any other descriptor or from a zero-terminated string. They can also be created empty and filled later since, although the data is non-modifiable, the entire contents of the buffer may be replaced by calling the assignment operator defined by the class. The replacement data may be another non-modifiable descriptor or a zero-terminated string, but in each case the new data length must not exceed the length specified in the template parameter when the buffer was created.

```
_LIT(KPalindrome, "Satan, oscillate my metallic sonatas");

TBufC<50> buf1(KPalindrome); // Constructed from literal descriptor
TBufC<50> buf2(buf1);        // Constructed from buf1
// Constructed from a NULL-terminated C string
TBufC<30> buf3((TText*)"Never odd or even");
TBufC<50> buf4;              // Constructed empty, length = 0

// Copy and replace
buf4 = buf1; // buf4 contains data copied from buf1, length modified
buf1 = buf3; // buf1 contains data copied from buf3, length modified
buf3 = buf2; // Panic! Max length of buf3 is insufficient for buf2 data
```

The class also defines a Des() method which returns a modifiable pointer descriptor for the data represented by the buffer. So, while the content of a non-modifiable buffer descriptor cannot normally be altered directly, other than by complete replacement of the data, it is possible to change the data indirectly by creating a modifiable pointer descriptor into the buffer. When the data is modified through the pointer descriptor, the lengths of both the pointer descriptor and the constant buffer descriptor are changed although, of course, the length is not automatically extended because the descriptor classes do not provide memory management.

```
_LIT8(KPalindrome, "Satan, oscillate my metallic sonatas");
TBufC8<40> buf(KPalindrome); // Constructed from literal descriptor
TPtr8 ptr(buf.Des()); // data is the string in buf, max length = 40

// Illustrates the use of ptr to copy and replace contents of buf
ptr = (TText8*)"Do Geese see God?";
```

```
ASSERT(ptr.Length()==buf.Length());
_LIT8(KPalindrome2, "Are we not drawn onward, we few, drawn onward to
      new era?");
ptr = KPalindrome2; // Panic! KPalindrome2 exceeds max length of ptr(=40)
```

TBuf<n>

Like the corresponding non-modifiable buffer class, this class for non-constant buffer data is a thin template class, the integer value determining the maximum allowed length of the buffer. It derives from `TBufBase`, which itself derives from `TDes`, thus inheriting the full range of descriptor operations in `TDes` and `TDesC`. `TBuf<n>` defines a number of constructors and assignment operators, similar to those offered by its non-modifiable counterpart.

As with all descriptors, its memory management is your responsibility and, although the buffer is modifiable, it cannot be extended beyond the initial maximum length set on construction. If the contents of the buffer need to expand, it's up to you to make sure that you either make the original allocation large enough at compile time or dynamically allocate the descriptor at runtime. The only way you can do the latter is to use a heap descriptor, described below. If this responsibility is too onerous and you want the resizing done for you, I suggest you consider using a dynamic buffer, as described in Chapter 7, bearing in mind that the associated overhead will be higher.

```
_LIT(KPalindrome, "Satan, oscillate my metallic sonatas");
TBuf<40> buf1(KPalindrome); // Constructed from literal descriptor
TBuf<40> buf2(buf1);        // Constructed from constant buffer descriptor
TBuf8<40> buf3((TText8*)"Do Geese see God?");  // from C string
TBuf<40> buf4; // Constructed empty, length = 0, maximum length = 40

// Illustrate copy and replace
buf4 = buf2; // buf2 copied into buf4, updating length and max length
buf3 = (TText8*)"Murder for a jar of red rum"; // updated from C string
```

The stack-based buffer descriptors, `TBuf<n>` and `TBufC<n>`, are useful for fixed-size or relatively small strings, say up to the length of a 256-character filename.

5.5 Heap-Based Buffer Descriptors

Heap-based descriptors can be used for string data that isn't in ROM and is not placed on the stack because it is too big. Heap-based descriptors can be used where they may have a longer lifetime than their creator, for example, passed to an asynchronous function. They are also useful where the length of a buffer to be created is not known at compile time and are used where malloc'd data would be used in C.

The class representing these descriptors is HBufC (explicitly HBufC8 or HBufC16) although these descriptors are always referred to by pointer, HBufC*. You'll notice that, by starting with "H", the class doesn't comply with the naming conventions described in Chapter 1. The class is exceptional and doesn't really fit any of the standard Symbian OS types exactly. Thus, it is simply prefixed with H to indicate that the data is stored on the heap. If you're interested in how the cleanup stack handles HBufC, see Chapter 3.

The HBufC class exports a number of static NewL() functions to create the buffer on the heap. These follow the two-phase construction model (described in Chapter 4) since they may leave if there is insufficient memory available. There are no public constructors and all heap buffers must be constructed using one of these methods (or from one of the Alloc() or AllocL() methods of the TDesC class which you may use to spawn an HBufC copy of any descriptor).

As you'll note from the "C" in the class name, these descriptors are non-modifiable, although, in common with the stack-based non-modifiable buffer descriptors, the class provides a set of assignment operators to allow the entire contents of the buffer to be replaced. As with TBufC, the length of the replacing data must not exceed the length of the heap cell allocated for the buffer or a panic occurs.

In common with TBufC, the heap-based descriptors can be manipulated at runtime by creating a modifiable pointer descriptor, TPtr, using the Des() method.

```
_LIT(KPalindrome, "Do Geese see God?");
TBufC<20> stackBuf(KPalindrome);

// Allocates an empty heap descriptor of max length 20
HBufC* heapBuf = HBufC::NewLC(20);
TInt length = heapBuf->Length();// Current length = 0
TPtr ptr(heapBuf->Des());        // Modification of the heap descriptor
ptr = stackBuf; // Copies stackBuf contents into heapBuf
length = heapBuf->Length();      // length = 17

HBufC* heapBuf2 = stackBuf.AllocLC(); // From stack buffer
length = heapBuf2->Length();            // length = 17

_LIT(KPalindrome2, "Palindrome");
*heapBuf2 = KPalindrome2;      // Copy and replace data in heapBuf2
length = heapBuf2->Length(); // length = 10
CleanupStack::PopAndDestroy(2, heapBuf);
```

Remember, the heap descriptors can be created dynamically to the size you require, but they are not automatically resized should you want to grow the buffer. You must ensure that the buffer has sufficient memory available for the modification operation you intend to use.

To help you with this, the HBufC class also defines a set of ReAllocL() methods to allow you to extend the heap buffer, which may potentially move the buffer from its previous location in memory (and may of course leave if there is insufficient memory). If the HBufC* is stored on the cleanup stack, moving the pointer as a result of memory reallocation can cause significant problems either in the event of a leave or if the cleanup stack's PopAndDestroy() function is used to destroy the memory. If you call Des() on a heap descriptor to acquire a TPtr, the iPtr member of this object is not guaranteed to be valid after re-allocation; it's safest to assume that the buffer will have moved and create a new TPtr accordingly.

There is no modifiable heap descriptor, HBuf, which you may have expected in order to make heap buffers symmetrical with TBuf stack buffers. The reasons for this are manifold. First, the expectation might reasonably be that the maximum length of a modifiable HBuf class would expand and contract on the heap as the content was modified. The goal of descriptors is efficiency, with memory managed by the programmer not the descriptor object, but an HBuf class which is modifiable but does not dynamically resize may be considered somewhat odd. To add dynamic resizing to HBuf would be difficult and costly in terms of code size because, as I've described, all the modifiable descriptor operations are implemented in its base class, TDes. In addition, if dynamic reallocation were added to HBuf, the programmer would have to be made aware that the location of the buffer might change as code is reallocated, and that certain functions might fail if there is insufficient memory, probably requiring a new set of leaving functions to be added to the class.

Heap buffers were initially intended to allow efficient reading of constant resource strings of variable length. Being non-modifiable, they save on the additional four bytes required to store a maximum length, which allows them to be as compact as possible. The Des() method permits them to be modified if necessary, when the programmer makes an explicit choice to do so. Providing a separate HBuf class would require the programmer to decide whether the buffer would ever require modification. It is possible that many would opt to use HBuf rather than HBufC "just in case" the descriptor later needed modification. Besides the additional code required for an HBuf class, the extra 4-byte overhead of HBuf objects created unnecessarily might add up over time. Providing the Des() method on the HBufC allows the programmer to modify heap buffers when needed but keeps the code associated with heap descriptor objects compact.

A counter-argument to this might be that the HBufC::Des() call is non-trivial and the resultant TPtr will occupy 12 bytes of stack space. In cases where a modifiable heap descriptor is definitely needed, creating a non-modifiable buffer and an additional, modifiable pointer to update it may be seen as a waste of processor instructions and memory. Perhaps the solution in future will be to provide both non-modifiable and modifiable heap buffers, with clear documentation as to the consequences of using each. With education, we should be able to trust developers to make the right choice of descriptor for the right operation, using constant heap descriptors by default and modifiable heap descriptors where necessary. I hope this book, and this chapter in particular, will go some way in helping to reach this goal!

To summarize, the inheritance hierarchy of the descriptor classes is shown in Figure 5.3.

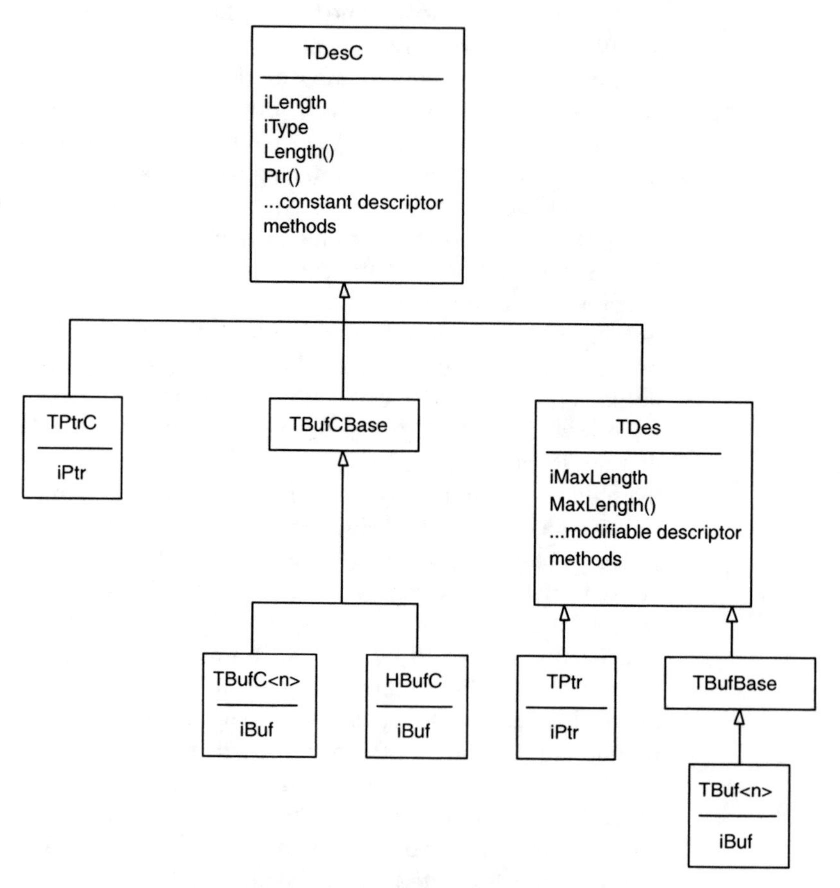

Figure 5.3 Class inheritance hierarchies

Heap descriptors can be created dynamically to the size you require, but are not automatically resized should they need to be extended beyond their maximum length.

5.6 Literal Descriptors

Let's move on to take a look at literal descriptors (constant descriptors that are compiled into ROM), which are equivalent to `static char[]` in C. Literal descriptors can be created by a set of macros defined in `e32def.h`. It's a bit unnerving at first sight, so I've only included the explicit definitions for 8- and 16-bit literals. The implicit definitions for the neutral macros `_L`, `_S` and `_LIT` are exactly the same, where `_L` is equivalent to `_L16` on a Unicode build and `_L8` on a narrow ASCII build:

```
#define _L8(a) (TPtrC8((const TText8 *)(a)))
#define _S8(a) ((const TText8 *)a)
#define _LIT8(name,s) const static TLitC8<sizeof(s)>
      name ={sizeof(s)-1,s}
#define _L16(a) (TPtrC16((const TText16 *)L ## a))
#define _S16(a) ((const TText16 *)L ## a)
#define _LIT16(name,s) const static TLitC16<sizeof(L##s)/2>
      name ={sizeof(L##s)/2-1,L##s}
```

Don't worry; I'll go through these slowly. Let's look at `_LIT` macros first, since these are the most efficient, and preferred, Symbian OS literals. The typical use of the macro would be as follows:

```
_LIT(KMyLiteralDescriptor, "The quick brown fox jumps over the lazy dog");
```

`KMyLiteralDescriptor` can then be used as a constant descriptor, for example written to a file or displayed to a user. The `_LIT` macro builds a named object (`KMyLiteralDescriptor`) of type `TLitC16` into the program binary, storing the appropriate string (in this case, *The quick brown fox jumps over the lazy dog*). As you'd expect, `_LIT8` and `_LIT16` behave similarly. The reason why the macros subtract one byte for the length of the data and divide by two, in the case of `_LIT16`, is that the macro is converting the C byte string to data which can be used as a descriptor.

For reference, here's the definition of class `TLitC16`, from `e32des.h` and `e32des.inl`, where `__TText` is typedef'd to a wide, 16-bit character. The `TLitC8` class, which has an array of 8-bit characters, has a similar definition.

```
template <TInt S>
class TLitC16
    {
public:
    inline const TDesC16* operator&() const;
    inline operator const TDesC16&() const;
    inline const TDesC16& operator()() const;
    ... // Omitted for clarity
```

```
public:
    TUint iTypeLength;
    __TText iBuf[__Align16(S)];
    };

template <TInt S>
inline const TDesC16* TLitC16<S>::operator&() const
        {return REINTERPRET_CAST(const TDesC16*,this);}
template <TInt S>
inline const TDesC16& TLitC16<S>::operator()() const
        {return *operator&();}
template <TInt S>
inline TLitC16<S>::operator const TDesC16&() const
        {return *operator&();}
```

As you can see, `TLitC16` (and `TLitC8`) do not derive from `TDesC8` or `TDesC16` but they have the same binary layouts as `TBufC8` or `TBufC16`. This allows objects of these types to be used wherever `TDesC` is used.[4]

You can form a pointer descriptor from the literal as follows:

```
TPtrC8 thePtr(KMyLiteralDescriptor);
```

It's slightly trickier to form a buffer descriptor from the literal. If you use `sizeof()` on a `_LIT` constant, the **size** of the corresponding `TLitC` object is returned, which is the size of the descriptor contents – in bytes – plus 8 extra bytes (a `TUint` for the stored length and the NULL terminator). If you want to use a stack-based buffer, you must take these extra bytes into account.

For a heap buffer, you can use the actual **length** of the descriptor contents to allocate the buffer then copy the contents of the descriptor. To get the correct length you can use the public `iTypeLength` member variable, or, more simply, use `operator()()` to `reinterpret_cast` the literal object to a descriptor and use the resultant object to determine the length of the contents. However, the simplest technique is to use `operator()()` to cast the literal object to a descriptor, then call one of the `TDes::AllocL()` methods upon it. For example:

```
// Define a 44 character literal
_LIT8(KExampleLit8, "The quick brown fox jumped over the lazy dog");

TInt size = sizeof(KExampleLit8); // 52 bytes (contents + 8 bytes)
TInt descriptorLength = KExampleLit8.iTypeLength; // 44 bytes

// Form a stack buffer descriptor around the literal
```

[4] Actually, the string stored in the program binary has a NULL terminator because the native compiler string is used to build it. However, as I described above, the length is adjusted to the correct value for a non-terminated descriptor by the `_LIT` macro as it constructs the `TLitC` object.

```
TBufC8<(sizeof(KExampleLit8)-8)> theStackBuffer(KExampleLit8);
// Create a heap buffer copying the contents of the literal
HBufC8* theHeapBuffer = KExampleLit8().AllocL();
// Similar behaviour for wide literals
_LIT16(KExampleLit16, "The quick brown fox jumped over the lazy dog");
size = sizeof(KExampleLit16);// 96 bytes (contents in bytes + 8 bytes)
descriptorLength = KExampleLit16.iTypeLength; // 44 bytes (contents)
```

Figure 5.4 illustrates the difference between the memory layouts for literal descriptors created using `_L` and `_LIT`.

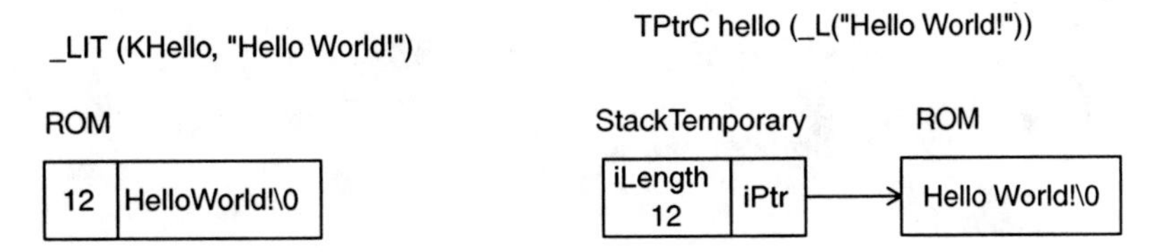

Figure 5.4 Memory layout for literal descriptors

Incidentally, literals have already been defined in Symbian OS to represent a blank string. There are three variants of the "NULL descriptor", defined as follows:

```
Build independent:               _LIT(KNullDesC,"");
8-bit for non-Unicode strings:   _LIT8(KNullDesC8,"");
16-bit for Unicode strings:      _LIT16(KNullDesC16,"");
```

Let's move on to look briefly at the `_L` and `_S` macros, the use of which is now deprecated in production code, though they may still be used in test code (where memory use is less critical). The advantage of using `_L` (or the explicit forms `_L8` and `_L16`) is that you can use it in place of a `TPtrC` without having to declare it separately from where it is used (besides saving an extra line of code, one benefit is that you don't have to think up a name for it!).

```
User::Panic(_L("example.dll"), KErrNotSupported);
```

The string ("`example.dll`") is built into the program binary as a basic, NULL-terminated string, with no initial length member (unlike the `TLitC` built for the `_LIT` macro). Because there is no length word, the layout of the stored literal is not like that of a descriptor and, when the code executes, each instance of `_L` will result in construction of a temporary `TPtrC`, with the pointer set to the address of the first byte of the literal as it is stored in ROM. The use of such a run-time temporary is safe as long as it is used only during the lifetime of the function in which it is created, or if it is copied for use outside of that lifetime.[5] However, the construction of a temporary, which requires setting the pointer, the length and the descriptor type, is an overhead in terms of inline constructor code which may bloat binaries where many string literals are used.

The `_S` macro is equivalent to the `_L` macro in terms of the way the string is stored in the binary, but it does not construct the temporary `TPtrC` around the string. These macros are useful if you wish to use the literal directly as a NULL-terminated string; you will incur no overhead if you do so.

Prefer the use of `_LIT` to `_L` for declaring literal descriptors, because the latter has an overhead associated with constructing a run-time temporary `TPtrC`.

5.7 Summary

This chapter introduced Symbian OS descriptors and discussed the following:

- The descriptor model treats string and binary data in the same way because there is no reliance on trailing NULL terminators to indicate the length of string data.
- The descriptor classes offer native "wide" 16-bit character support, but the same APIs can be used explicitly with 8-bit binary or string data.
- Non-modifiable descriptor functionality is defined by the `TDesC` base class and inherited by all its subclasses.
- For compactness, no virtual functions are used in the descriptor hierarchies, so no additional 4-byte `vptr` is added to each descriptor object.
- The base class, `TDesC`, defines `Length()` and uses the bottom 28 bits of the first machine word of the object to hold the length of the descriptor data. `TDesC` also defines `Ptr()` to access that data – it uses hardcoded logic to determine the correct memory address, based on the descriptor subtype which is indicated by the top 4 bits of the first machine word of the object.

[5] Code like this will not work in a DLL because, although marked `const`, the creation of such a runtime temporary constitutes the use of modifiable static data, which is disallowed on Symbian OS (as described in Chapter 13).

```
const TPtrC KDefaultPath=_L("C:\\System\\Messages")
```

- Modifiable descriptor functionality is defined by the `TDes` class, which derives from `TDesC`, adding an extra machine word to store the maximum allowed length of a descriptor. The `TDes` class implements typical data modification operations.
- The five concrete descriptor classes can be subdivided in terms of their memory layout (pointer or buffer), whether they are constant or modifiable and whether they are heap- or stack-based. However, there is significant interoperability between the classes, and the base class APIs make no assumptions about their underlying data layout. Descriptor data can be stored in RAM (on the stack or heap) or ROM; the APIs are consistent regardless of location or memory layout.
- For efficiency reasons, descriptors do not dynamically extend the data area they reference. You, the programmer, are responsible for memory management.
- Descriptor member functions check that access to the descriptor lies within its data area and raise a panic (in both debug and release builds) if a descriptor overflow would result. This ensures that descriptor code is robust and that programming errors are easy to track down.
- `_LIT` literals are not part of the `TDesC` inheritance hierarchy, but have an equivalent memory layout in ROM to `TBufC` and can thus be used interchangeably.
- The advantage of using the `_L` literal macros is that you can use them where you would use a temporary `TPtrC`, without having to predefine and name them. The disadvantage is the extra run-time overhead associated with construction of a temporary descriptor – which is why they are deprecated in production code.
- Symbian OS descriptors may take some getting used to, but cannot be avoided when programming for Symbian OS because many API functions use them.

The next chapter shows how to use descriptors effectively and examines the more frequently used descriptor functions. It also describes some common descriptor mistakes and misconceptions.

6

Good Descriptor Style

For every problem there is one solution which is simple, neat and wrong

H L Mencken

The previous chapter covered the basics of Symbian OS strings, known as descriptors. It examined the methods used to instantiate each concrete descriptor class and described how to access, modify and replace the descriptor data. It also discussed the descriptor base classes (the non-modifiable base class `TDesC`, which implements constant descriptor operations, and `TDes`, which derives from it and implements methods for descriptor modification). The chapter should have given you a good idea of the concrete types of descriptor.

This chapter examines some of the descriptor manipulation methods of the base classes and discusses mistakes and problems commonly encountered when using descriptors. Figure 6.1 summarizes the descriptor classes and the factors to bear in mind when deciding which type of descriptor to use.

6.1 Descriptors as Parameters and Return Types

When writing code, you probably don't want to be constrained to using a `TBuf` just because a particular library function requires it. Likewise, as a function provider, you're probably not interested in the type of descriptor your callers will be passing to you. In fact, you shouldn't require a particular type, because if you change the implementation later, you may want to change the type of descriptor, and if you expose it at the API level you will require your clients to change their code. This kind of change breaks source compatibility and is highly undesirable. I'll discuss compatibility further in Chapter 18.

Unless you're taking ownership, you don't even need to know if the incoming descriptor parameter or return value is stack- or heap-based. In fact, as long as the descriptor is one of the standard types, so the

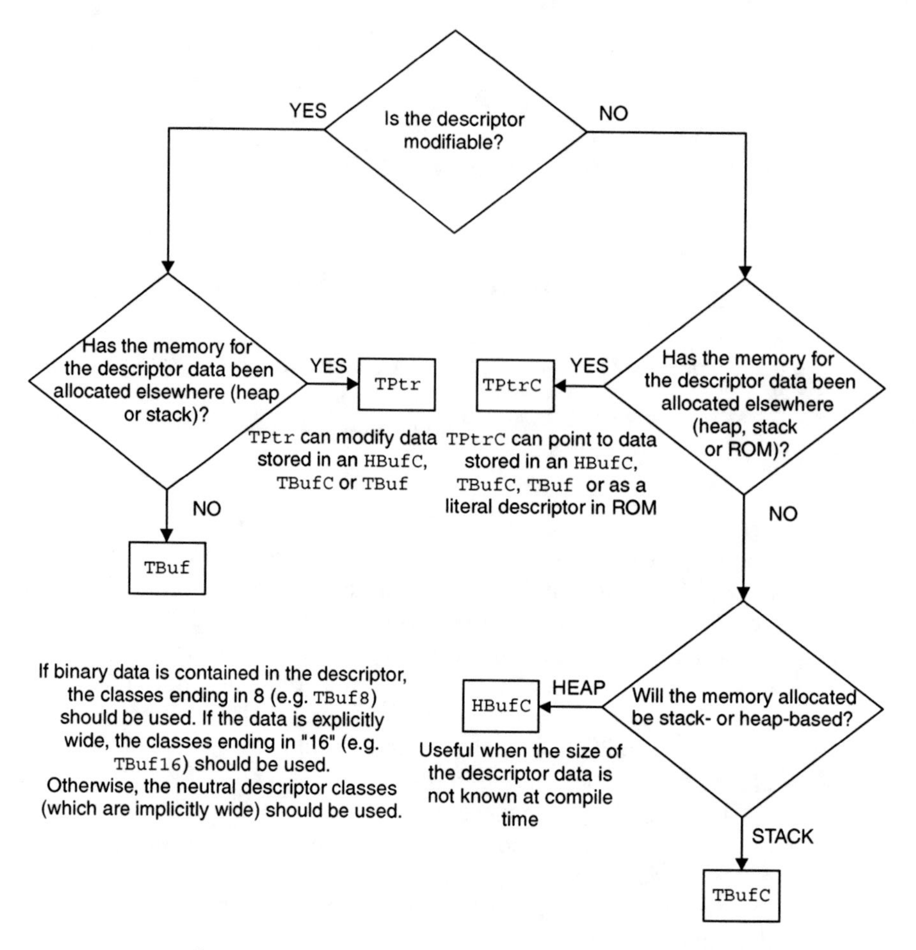

Figure 6.1 Flow chart to choose the correct descriptor type

appropriate descriptor methods can be called on it, the receiving code can remain blissfully ignorant as to its layout and location in memory. For this reason, when defining functions you should always use the abstract base classes as parameters or return values. For efficiency, descriptor parameters should be passed by reference, either as `const TDesC&` for constant descriptors or `TDes&` when modifiable. I'll discuss good API design in detail in Chapter 20.

As an example, class `RFile` defines straightforward file read and write methods as follows:

```
IMPORT_C TInt Write(const TDesC8& aDes);
IMPORT_C TInt Read(TDes8& aDes) const;
```

For both methods, the input descriptor is explicitly 8-bit to allow for both string and binary data within a file. The descriptor to write to the file is a constant reference to a non-modifiable descriptor, while to read from the file requires a modifiable descriptor. The maximum length of the modifiable descriptor determines how much file data can be read into it, so the file server doesn't need to be passed a separate parameter to describe the length of the available space. The file server fills the descriptor unless the content of the file is shorter than the maximum length, in which case it writes what is available into the descriptor. The resultant length of the descriptor thus reflects the amount of data written into it, so the caller does not need to pass a separate parameter to determine the length of data returned.[1]

When writing a function which receives a modifiable descriptor you don't actually need to know whether it has sufficient memory allocated to it, since the descriptor methods themselves perform bounds checking and panic if the operation would overflow the descriptor. Of course, you may not always want the descriptor methods to panic if the caller's descriptor data area is too short. Sometimes the caller cannot know until runtime the maximum length required and a panic is somewhat terminal. You should define clearly in your documentation what happens when the descriptor is not large enough. There are times when a more suitable approach would be to return the length required to the caller so it can take appropriate steps to allocate a descriptor of the correct length. For example:

```
HBufC* CPoem::DoGetLineL(TInt aLineNumber)
   {// Code omitted for clarity. Allocates and returns a heap buffer
   // containing the text of aLineNumber (leaves if aLineNumber is
   // out of range)
   }

void CPoem::GetLineL(TInt aLineNumber, TDes& aDes)
   {
   HBufC* line = DoGetLineL(aLineNumber);
   CleanupStack::PushL(line);

   // Is the descriptor large enough (4 bytes or more) to return an
   // integer representing the length of data required?
   if (aDes.MaxLength() < line->Length())
      {
      if (aDes.MaxLength() >= sizeof(TInt))
         {// Writes the length required (TPckg is described later)
         TPckg<TInt> length(line->Length());
         aDes.Copy(length);
         }

      // Leave & indicate that the current length is too short
```

[1] For comparison purposes only, here's a similar function from the Win32 Platform SDK, which uses a basic buffer (lpBuffer) for file reads and thus requires extra parameters to indicate the amount to read and the amount that was actually read:

```
BOOL ReadFile(HANDLE hFile, LPVOID lpBuffer,
        DWORD nNumberofBytesToRead, LPDWORD lpNumberofBytesRead,
        LPOVERLAPPED lpOverlapped);
```

```
        User::Leave(KErrOverflow); // Leaves are described in Chapter 2
        }
    else
        {
        aDes.Copy(*line);
        CleanupStack::PopAndDestroy(line);
        }
    }
```

An alternative approach is to allocate a heap buffer of the appropriate size for the operation and return it to the caller. The caller takes ownership of the heap buffer on return and is responsible for deleting it:

```
HBufC8* CPoem::GetLineL(TInt aLineNumber)
    {
    return (DoGetLineL(aLineNumber)); // As shown above
    }

void PrintPoemLinesL(CPoem& aPoem)
    {
    HBufC* line;
    FOREVER // See footnote 2
        {
        line = poem.NextLineL();
        ... // Do something with line (make it leave-safe if necessary)
        delete line;
        }
    }
```

When defining functions, use the abstract base classes as parameters and return values. For efficiency, pass them by reference, either as `const TDesC&` for constant descriptors or `TDes&` when modifiable.

6.2 Common Descriptor Methods

Moving on, let's consider some of the methods implemented by the descriptor base classes, particularly those that may cause problems if misused. The SDK documentation covers the descriptor classes extensively, and you'll find more information on each of the descriptor methods I

[2] The `FOREVER` macro is defined in `e32def.h` as follows:

```
#define FOREVER for(;;)
```

The code will run in the loop until the end of the poem is reached. At that point `NextLineL()` will leave because `CPoem::DoGetNextLineL()` leaves to indicate the end of the poem. This breaks the code out of the loop, so it will not run forever, only until the end of the poem. Some poems, of course, do seem to go on forever...

discuss here, as well as others for seeking, comparison, extraction, string and character manipulation and formatting.

The base classes implement methods to access and manipulate the string data of deriving classes. The base classes do not themselves store the data. They have no public constructors save an implicit copy constructor. So you won't attempt to instantiate them. Will you?

The `TDesC` base class implements `Ptr()` to provide access to the descriptor data, as described in Chapter 5. The method returns a pointer to the first character in the data array, allowing you to manipulate the descriptor contents directly, using C pointers, if you so wish.[3] In doing so, you should take care not to write off the end of the descriptor data area. (One way to check that this kind of programming error does not occur is to use assertion statements, as described in detail in Chapter 16.) I've included an example of using a pointer to manipulate descriptor data later in this chapter, to illustrate an issue related to the maximum length of heap descriptors.

`TDesC` implements both `Size()` and `Length()` methods, and you should be careful to differentiate between them. The `Size()` method returns the number of bytes the descriptor occupies, while the `Length()` method returns the number of characters it contains. For 8-bit descriptors, this is the same thing, i.e. the size of a character is a byte. However, from Symbian OS v5u, the native character is 16 bits wide, which means that each character occupies two bytes. For this reason, `Size()` always returns a value double that of `Length()`. (As I described in Chapter 5, the length of a descriptor cannot be greater than 2^{28} bytes because the top four bits of the length word are reserved to indicate the type of descriptor.)

The base class for modifiable descriptors, `TDes`, implements a `MaxLength()` method which returns the maximum length allowed for the descriptor. Beware of the `SetMax()` method, which doesn't allow you to change the maximum length of the descriptor, thereby expanding or contracting the data area; instead, it sets the current length of the descriptor to the maximum length allowed.

You can use `SetLength()` to adjust the descriptor length to any value between zero and its maximum length (inclusively). The `Zero()` method also allows you to set the length to zero. If you thought that method might fill the contents of the descriptor with zeroes, you want the `FillZ()` method – which is overloaded so you can fill the entire descriptor or to a required length. If you don't want to fill with zeroes, the overloaded `Fill()` methods permit you to choose which character to fill the descriptor.

---

[3] `TDes` also has the `PtrZ()` method, which returns a pointer to the first character in the data array and appends a NULL terminator to the descriptor data so the returned pointer can be used directly as a C string. (The length of the descriptor must be less than its maximum allowable length in order for there to be sufficient space for the zero terminator.)

TDes implements an overloaded set of Copy() methods, two of which are shown below. (There are another six overloads of the Copy() function, which allow copying directly from a NULL-terminated string or from a pointer directly into descriptor data and which perform folding, collation or case adjustment as part of the copy.)

```
IMPORT_C void Copy(const TDesC8 &aDes);
IMPORT_C void Copy(const TDesC16 &aDes);
```

These methods copy the descriptor parameter into the data area of the descriptor on which they are called, setting its length to reflect the new data. The methods will panic if the maximum length of the receiving descriptor is shorter than the incoming data. The Copy() method is overloaded to take either an 8- or 16-bit descriptor. Thus, not only is it possible to copy a narrow-width descriptor onto a narrow-width descriptor and a wide descriptor onto a wide descriptor, but it is also possible to copy between descriptor widths, effecting a conversion of sorts in the process.

These methods are an example of where the implementation of the TDes8 class differs from the TDes16 class. As Figure 6.2 illustrates, the Copy() method implemented by TDes8 to copy an incoming wide 16-bit descriptor into a narrow descriptor *strips out alternate characters,* assuming them to be zeroes, that is, the data values should not exceed 255 (decimal). The Copy() method which copies a narrow descriptor into the data area is a straight data copy, however.

```
// Instantiate a narrow descriptor
TBuf8<3> cat(_L8("cat")); // _L is described in Chapter 5
// Instantiate a wide descriptor
TBuf16<3> dog(_L16("dog"));
// Copy the contents of the wide descriptor into the narrow descriptor
cat.Copy(dog); // cat now contains "dog"
```

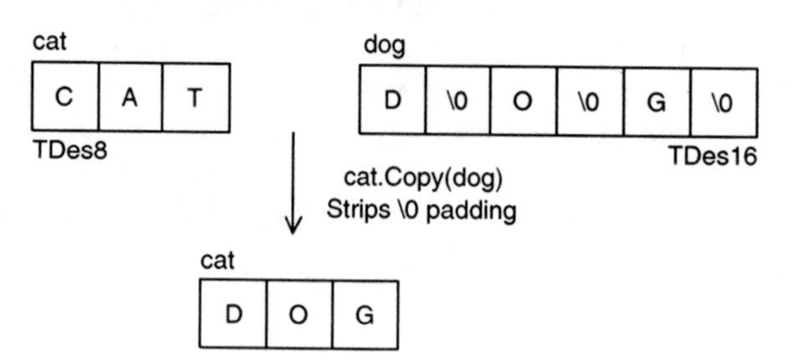

Figure 6.2 NULL characters are stripped out when wide characters are copied into a narrow descriptor

Conversely, for class `TDes16`, an incoming 16-bit descriptor can be copied directly onto the data area, but the `Copy()` method that takes an 8-bit descriptor *pads each character with a trailing zero* as part of the copy operation, as shown in Figure 6.3. The `Copy()` methods thus form a rudimentary means of copying and converting when the character set is encoded by one 8-bit byte per character and the last byte of each wide character is simply a NULL character padding.

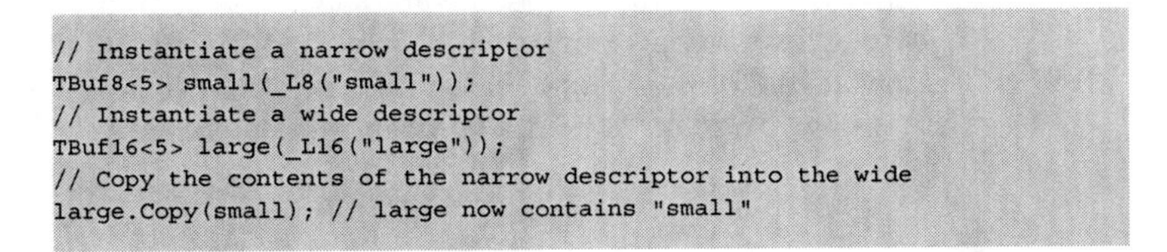

```
// Instantiate a narrow descriptor
TBuf8<5> small(_L8("small"));
// Instantiate a wide descriptor
TBuf16<5> large(_L16("large"));
// Copy the contents of the narrow descriptor into the wide
large.Copy(small); // large now contains "small"
```

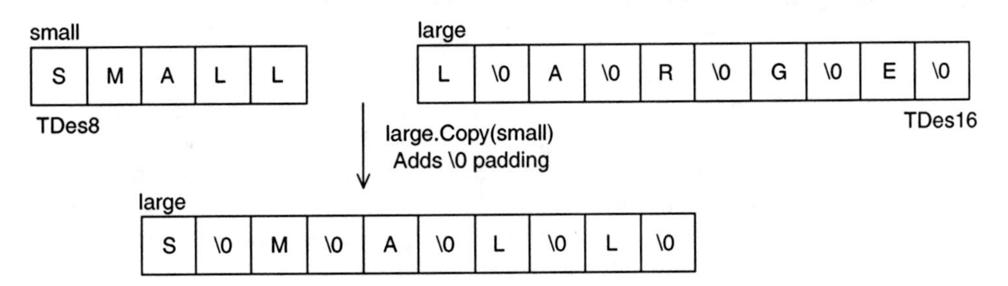

Figure 6.3 NULL characters pad narrow characters when copied into a wide descriptor

To perform proper conversion in both directions between 16-bit Unicode and 8-bit, non-Unicode, character sets (or between Unicode and the UTF-7 and UTF-8 transformation sets) you should use the conversion library (`charconv.lib`) that Symbian OS provides (see header file `charconv.h`). You can also use the Character Conversion Plug-in Provider API to write plug-in DLLs to extend the range of foreign character sets beyond the ones already provided by Symbian OS. I'll discuss the use of a framework and plug-ins, a commonly used Symbian OS idiom, in more detail in Chapter 13.

Symbian OS has been built as `_UNICODE` since the v5U release. The descriptor classes which end in 8 or 16 reflect the character size of the descriptor. The neutral versions, with no numerical indicator, are implicitly 16-bit and, for many descriptor operations, you can simply use this neutral version, which will, if nothing else, make your code more readable. However, there are occasions when you are dealing with text of a specific character size. For example, you may be working with 8-bit text (e.g. an ASCII text file or Internet email – or simply using the `RFile::Read()` and `RFile::Write()`

methods shown above) for which you should explicitly use the 8-bit descriptor classes, using TText8 pointers as returned by the Ptr() operation to access characters. When working with 16-bit text (for example, Java strings), you should indicate the fact explicitly by using the TDesC16-derived classes, referencing individual characters with TText16 pointers.

As I described in Chapter 5, because descriptors do not use NULL terminators, they may be used with binary data as well as strings, which also allows code re-use within Symbian OS. Binary data is always considered to be 8-bit and you should use the 8-bit classes to manipulate it, using TInt8 or TUint8 to reference individual bytes.

6.3 The Use of HBufC Heap Descriptors

Having discussed some of the features of the descriptor classes, I'll move on now to discuss some of the common mistakes made when using descriptors. First I'll cover the creation and use of HBufC heap descriptors.

As I mentioned in Chapter 5, HBufC can be spawned from existing descriptors using the Alloc() or AllocL() overloads implemented by TDesC. Here is a contrived example which shows how to replace inefficient code with AllocL():

```
void CSampleClass::UnnecessaryCodeL(const TDesC& aDes)
    {
    iHeapBuffer = HBufC::NewL(aDes.Length());
    TPtr ptr(iHeapBuffer->Des());
    ptr.Copy(aDes);
    ...
    // could be replaced by a single line
    iHeapBuffer = aDes.AllocL();
    }
```

Another common way to introduce complexity occurs in the opposite direction, that is, the generation of TDesC& from a heap descriptor. A common mistake is to call the Des() method on the heap descriptor; this is not incorrect (it returns a TDes&), but it is clearer and more efficient simply to de-reference the HBufC pointer when a non-modifiable TDesC& is required:

```
const TDesC& CSampleClass::MoreAccidentalComplexity()
    {
    return (iHeapBuffer->Des());
    // could be replaced more efficiently with
    return (*iHeapBuffer);
    }
```

Another subtle problem occurs when you allocate an HBufC and then call Des() on it to return a TPtr. If you recall, an HBufC object doesn't have a maximum length word – since it is non-modifiable, it doesn't need one. But the modifiable TPtr does. The length of the TPtr is set to the stored length of the HBufC, but where does the maximum length come from when you create it? In fact, when you call Des() on HBufC, it uses the maximum length of the heap cell in which the HBufC was allocated to set the maximum length of the returned TPtr.

The maximum length of the heap cell may not be exactly what you specified as the maximum length of the heap descriptor when you allocated it. This may happen because you didn't specify a word-aligned maximum length (i.e. a multiple of 4 bytes) or because there was not enough space left over in the free cell from which the heap cell was allocated to create any other heap cells. The minimum size required for a heap cell is approximately 12 bytes and, if there are fewer bytes left over, your descriptor will be given the extra bytes too. (The former case of specifying an unaligned maximum length is much more common.) The end result in either case is that the maximum length of a TPtr returned from a call to HBufC::Des() may not be exactly the size you asked for when you allocated the heap descriptor; while it will not be truncated, it could be longer. Don't get caught out by the fact that it may be larger than you expect – but, likewise, don't expect that it is simply rounded up to the next word-aligned value. For example:

```
HBufC8* buf = HBufC8::NewLC(9);
TPtr8 ptr(buf->Des());
TInt maxLength = ptr.MaxLength(); // maxLength>9 but may not be 12
```

In practice this will be guaranteed to fill ptr with at least three extra bytes but beyond that, you cannot predict how much larger the maximum length of the heap buffer is than requested. Since you cannot guarantee the value of the maximum length, stick to using the Length() methods of the TPtr or HBufC, or the value you used to allocate the HBufC initially.

Here's an example where you could get caught out, illustrating the use of pointers to manipulate the contents of a descriptor and the use of an assertion statement to catch access beyond the descriptor length:

```
_LIT(KPanic, "TestPointer");
const TInt KBufferLength = 10;

void TestPointer()
    {// Create a buffer with length KBufferLength = 10 bytes
    HBufC8* myBuffer = HBufC8::NewMaxL(KBufferLength);

    TPtr8 myPtr(myBuffer->Des());
    myPtr.Fill('?'); // Fill with '?'
```

```
// Byte pointer to descriptor in memory
TUint8* ptr = (TUint8*)myPtr.Ptr();

TInt maxLength = myPtr.MaxLength();
for (TInt index = 0; index < maxLength; index++)
    {// This fails at the end of the buffer (index = 10)
    //  because myPtr.MaxLength() > KBufferLength
    __ASSERT_DEBUG(index<KBufferLength,
            User::Panic(KPanic, KErrOverflow));
    (*ptr) = '!'; // Replace the contents with '!'
    ++ptr;
    }
}
```

A common mistake is to call the `Des()` method on the heap descriptor to return a `TDes&`. It is more efficient simply to de-reference the `HBufC` pointer when a non-modifiable `TDesC&` is required.

6.4 Externalizing and Internalizing Descriptors

Moving on from the accidental complexity that may be introduced when using heap descriptors to a more general issue – that of externalizing descriptor data to file using a writable stream and re-internalizing it using a readable stream. Consider the following sample code:

```
// Writes the contents of iHeapBuffer to a writable stream
void CSampleClass::ExternalizeL(RWriteStream& aStream) const
    {
    // Write the descriptor's length
    aStream.WriteUint32L(iHeapBuffer->Length());
    // Write the descriptor's data
    aStream.WriteL(*iHeapBuffer, iHeapBuffer->Length());
    }

// Instantiates iHeapBuffer by reading the contents of the stream
void CSomeClass::InternalizeL(RReadStream& aStream)
    {
    TInt size=aStream.ReadUint32L(); // Read the descriptor's length
    iHeapBuffer = HBufC::NewL(size);  // Allocate iHeapBuffer
    // Create a modifiable descriptor over iHeapBuffer
    TPtr ptr(iHeapBuffer->Des());
    // Read the descriptor data into iHeapBuffer
    aStream.ReadL(ptr,size);
    }
```

The code above implements descriptor externalization and internalization in a very basic manner, using four bytes to store the descriptor length, then adding the descriptor data to the stream separately. In the `InternalizeL()` method, the descriptor is reconstructed in four rather awkward stages. In fact, Symbian OS provides a templated stream operator (`operator<<`) for externalization, which compresses the length information to keep descriptor storage as efficient and compact as possible. Furthermore, descriptors externalized in this way may be re-created from the stream, as heap descriptors using the `NewL()` overloads of `HBufC`, passing in the read stream and an additional parameter to indicate the maximum length of data to be read from the stream. This is a significantly more efficient approach. To use `operator<<` you must link to `estor.lib`. The stream classes are documented in your preferred SDK.

```
void CSampleClass::ExternalizeL(RWriteStream& aStream) const
    {// Much more efficient, no wasted storage space
    aStream << iHeapBuffer;
    }

void CSampleClass::InternalizeL(RReadStream& aStream)
    {
    iHeapBuffer = HBufC::NewL(aStream, KMaxLength);
    }
```

You can also use the templated `operator>>`, which assumes that the descriptor was externalized using `operator<<`.

```
class TSomeClass
    {
    ... // Omitted for clarity
private:
    TBuf<12> iBuffer;
    ...
    };

void TSomeClass::ExternalizeL(RWriteStream& aStream)
    {
    aStream << iBuffer;
    ...
    }

void TSomeClass::InternalizeL(RReadStream& aStream)
    {
    aStream >> iBuffer;
    ...
    }
```

You'll notice that the internalization and externalization methods above are leaving functions. This is because the templated streaming operators, `operator>>` and `operator<<`, may leave. They are not suffixed with `L`, as is conventional for all leaving functions, because they are operators, and to do so would interfere with their usage. The fact that they can leave isn't picked up by the leave-checking tool LeaveScan (described in Chapter 2) either. When writing code which uses them you should make a special point of checking that your code is leave-safe and that functions are named according to Symbian OS.

Symbian OS also provides `WriteL()` and `ReadL()` methods on the `RWriteStream` and `RReadStream` classes, respectively. The `WriteL()` method writes only the contents of the descriptor, not the length. The `ReadL()` method reads the contents into a target descriptor up to its maximum length.

Symbian OS provides templated stream operators for externalization and internalization of descriptors, to store the descriptor efficiently. These should be used in preference to hand-crafted code when reading and writing a descriptor to a stream. Where only the contents of the descriptor need to be stored, `ReadL()` and `WriteL()` should be used.

6.5 The Overuse of `TFileName`

Another potential hazard when using descriptors occurs through the overuse of `TFileName` objects. `TFileName` is defined as follows in `e32std.h`:

```
const TInt KMaxFileName=0x100; // = 256 (decimal)
typedef TBuf<KMaxFileName> TFileName;
```

Since each wide character is equivalent to two bytes, each time you create and use a `TFileName` object on the stack you are setting aside 524 bytes (2×256 descriptor data bytes $+ 12$ for the descriptor object itself) regardless of whether they are all actually required for the file name. The standard stack size on Symbian OS is just 8 KB, so it's a good rule never to allocate stack-based `TFileName` objects or pass them around by value, since unnecessary use of this restricted resource is very wasteful. You could, of course, use them on the heap, say as members of C classes (which always exist on the heap, as described in Chapter 1). But if you are unlikely to need the full path length, you should aim to use an `HBufC` or some other descriptor type, since it is good practice to consume as little memory as necessary.

As an example of where you can often avoid using an unnecessary `TFileName` object, consider the `TParse` class (defined in `f32file.h`). This class takes a copy of a descriptor containing a file name to parse, and stores it in a `TFileName`, which may waste valuable stack space. You should consider using the `TParsePtr` and `TParsePtrC` classes instead; these offer the same functionality (implemented by the base class `TParseBase`) while storing a reference to the file name rather than a copy.

Do not waste valuable stack resources by using TFileName or TParse unnecessarily.

6.6 Useful Classes for Descriptor Manipulation

Having looked at a few common problems, let's close this chapter by taking a quick look at a couple of useful classes you can use with descriptors.

Firstly, let's look at lexical analysis and extraction, using the TLex class. TLex is provided as TLex8 and TLex16 in the same way as descriptors, though you should use the build-independent type (implicitly TLex16) unless you require a particular build variant. The class implements general-purpose lexical analysis, and effects syntactical element parsing and string-to-number conversion, using the locale-dependent functions of TChar to determine whether each character is a digit, a letter or a symbol. An object of type TLex stores a pointer to the string data to be analyzed. In addition, the class maintains an extraction mark to indicate the current lexical element and a pointer to the next character to be examined. The marker can be used to mark position and you can peek, skip and rewind as required.

Another useful set of classes are the package buffers (TPckgBuf) and package pointers (TPckg and TPckgC) which are thin template classes derived from TBuf<n>, TPtr and TPtrC respectively (see e32std.h). The classes are type-safe and are templated on the type to be packaged – I'll discuss the thin template idiom later, in Chapter 19. The package classes allow flat data objects to be stored conveniently within descriptors, which is useful for example for inter-thread or inter-process data transfer (which I will describe further in Chapters 10, 11 and 12). In effect, a T class object may be packaged whole into a descriptor (I like to think of this as "descriptorizing") so it may be passed easily in a type-safe way between threads or processes.

There are two package pointer classes, creating either modifiable (TPckg) or non-modifiable (TPckgC) pointer descriptors which refer to the existing instance of the template-packaged class. Functions may be called on the enclosed object, although if it is enclosed in a TPckgC, a constant reference to the packaged object is returned from operator().

The package buffer TPckgBuf creates and stores a new instance of the type to be encapsulated in a descriptor. The copied object is owned by the package buffer; it is modifiable and functions may be called on it, after calling operator() on the TPckgBuf object to retrieve it. Because the package buffer contains a copy of the original object, if a modification function is called it is the copy that is modified – the original is unchanged.

The following code shows an object of a simple T class encapsulated in the package types, and Figure 6.4 illustrates the memory layout of each:

```
class TSample
    {
public:
    void SampleFunction();
    void ConstantSampleFunction() const;
private:
    TInt iSampleData;
    };

TSample theSample;
TPckg<TSample> packagePtr(theSample);
TPckgC<TSample> packagePtrC(theSample);
TPckgBuf<TSample> packageBuf(theSample);

packagePtr().SampleFunction();
packagePtrC().SampleFunction();// Compile error! Non-const function
packagePtrC().ConstantSampleFunction();
packageBuf().SampleFunction();
```

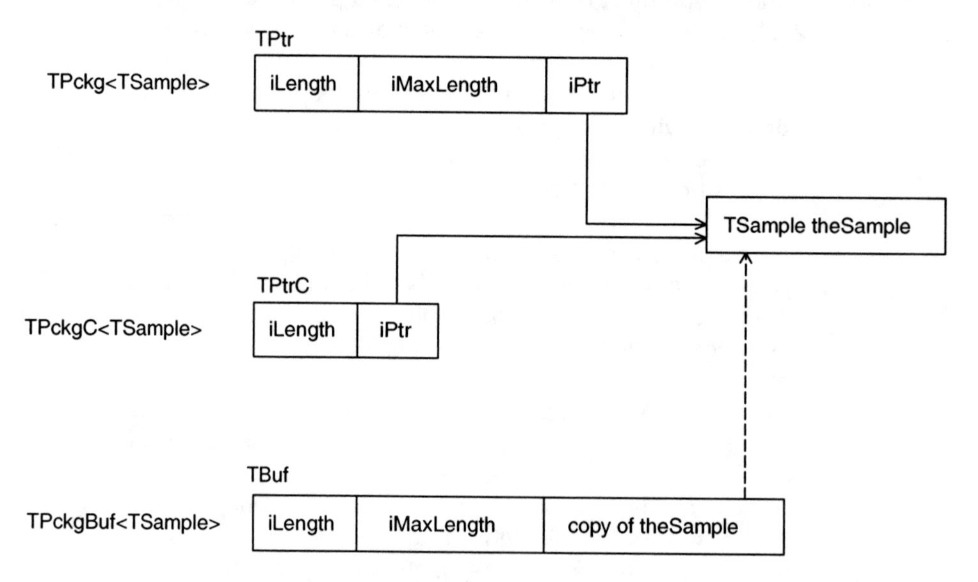

Figure 6.4 Memory layout of the `TPckg`, `TPckgC` and `TPckgBuf` classes

6.7 Summary

In this chapter, I discussed the following:

- You should not attempt to instantiate objects of the base descriptor classes `TDesC` and `TDes`. However, you will typically use them when defining descriptor function parameters because they give a caller flexibility as to the type of descriptor passed to the function. For example, the `Read()` and `Write()` methods of class `RFile` take a generic (8-bit) descriptor.

- One of the benefits of descriptor parameters is that the function can determine the length of the descriptor, and its maximum allowed length if it is modifiable. This means that a function does not require separate parameters for this information.
- `TDesC` defines a `Size()` method, which returns the total size in bytes of the current descriptor contents, and a `Length()` method, which returns the number of characters currently contained by the descriptor. From v5U, the native character on Symbian OS is 16 bits (two bytes) wide; as a result, `Size()` will always return a value double that of `Length()`.
- Conversion between 16-bit and 8-bit descriptors can be effected using the `Copy()` function defined by `TDes`. However, this mode of copying assumes basic character data (i.e. values not exceeding 255) because it performs rudimentary zero-padding for narrow to wide conversion and zero-stripping for wide to narrow conversion. Symbian OS provides the `charconv` library to convert between genuine 16-bit Unicode and 8-bit, non-Unicode character sets (or between Unicode and the UTF-7 and UTF-8 transformation sets).
- If you are working with 8-bit text or binary data you should explicitly use the 8-bit descriptor classes, using `TText8` pointers as returned by the `Ptr()` operation to access characters. Otherwise, you should use the neutral descriptor classes unless you wish to highlight the fact that the data is explicitly 16 bits wide.
- A number of common mistakes are made with `HBufC` heap descriptors, including:
  - using long-winded allocation from other descriptors rather than calling `TDesC::AllocL()` or `TDesC::AllocLC()`
  - using the `Des()` method to return a const `TDesC&` rather than simply dereferencing the `HBufC` pointer
  - calling `MaxLength()` on a heap descriptor, which can return a spurious value
- The templated `>>` and `<<` stream operators should be used to internalize and externalize descriptors using `RReadStream` and `RWriteStream` respectively; or alternatively, the `ReadL()` and `WriteL()` methods of the stream classes should be used. These have been optimized for the most efficient storage of descriptor data.

- Use of stack-based objects of types `TFileName` and `TParse` should be kept to a minimum because they reserve a large amount of stack space (524 bytes), often unnecessarily.
- Symbian OS provides the `TLex` classes for lexical analysis and extraction.
- The package descriptor classes `TPckgBuf`, `TPckg` and `TPckgC` are useful template classes for type-safe wrapping of the built-in types or T class objects within descriptors. This method of "descriptorizing" flat data is particularly useful for passing objects between client and server, and is described in detail in Chapters 10, 11 and 12.

7

Dynamic Arrays and Buffers

Make my skin into drumheads for the Bohemian cause

The last words of Czech General Jan Zizka (1358–1424)

This chapter discusses the use of the dynamic array classes `RArray<class T>` and `RPointerArray<class T>`. It also describes the `CArrayX<class T>` classes and the dynamic buffer classes they use. Dynamic array classes are very useful for manipulating collections of data without needing to know in advance how much memory to allocate for their storage. They expand as elements are added to them and, unlike C++ arrays, do not need to be created with a fixed size.

Conceptually, the logical layout of an array is linear, like a vector. However, the implementation of the dynamic array can either use a single heap cell as a "flat" buffer to hold the array elements or allocate the array buffer in a number of segments, using a doubly-linked list to manage the segmented heap memory. Contiguous flat buffers are typically used when high-speed pointer lookup is an important consideration and when array resizing is expected to be infrequent. Segmented buffers are preferable for large amounts of data and where the array is expected to resize frequently, or where a number of elements may be inserted into or deleted from the array.

The *capacity* of a dynamic array is the number of elements the array can hold within the space currently allocated to its buffer. When the capacity is filled, the array dynamically resizes itself by reallocating heap memory when the next element is added. The number of additional elements allocated to the buffer is determined by the *granularity*, which is specified at construction time.

It is important to choose an array granularity consistent with the expected usage pattern of the array. If too small a value is used, an overhead will be incurred for multiple extra allocations when a large number of elements are added to the array. However, if too large a granularity is chosen, the array will waste storage space.

For example, if an array typically holds 8 to 10 objects, then a granularity of 10 would be sensible. A granularity of 100 would be unnecessary. However, if there are usually 11 objects, a granularity of 10 wastes memory for 9 objects unnecessarily. A granularity of 1 would also be foolish, since it would incur multiple reallocations.

Symbian OS provides a number of different dynamic array classes with names prefixed by "CArray", such as CArrayFixFlat, CArrayFixSeg and CArrayVarSeg (which I'll refer to collectively as "CArrayX"), as well as the RArray and RPointerArray classes. It can be quite difficult to determine which to use, so this chapter guides you through their main characteristics.

Choose the granularity of a dynamic array carefully to avoid wasting storage space (if the granularity chosen is too large) or frequent re-allocations (if it is chosen too small).

7.1 CArrayX Classes

There are a number of CArrayX classes, which makes this dynamic array family very flexible, albeit with an associated performance overhead which I'll discuss later in this chapter. To sidestep the performance penalty, the RArray and RPointerArray classes were added to Symbian OS to provide simpler and more efficient dynamic arrays. You should use these classes in preference to CArrayX where possible.

However, for background information, I'll run through some brief details of the CArrayX classes. The naming scheme works as follows; for each class the CArray prefix is followed by:

- Fix for elements which have the same length and are copied so they may be contained in the array buffer.
- Var where the elements are of different lengths; each element is contained within its own heap cell and the array buffer contains *pointers* to the elements.
- Pak for a packed array where the elements are of variable length; they are copied so they may be contained within the array buffer. Each element is preceded by its length information, rather like a descriptor.
- Ptr for an array of pointers to CBase-derived objects.

Following this, the array class name ends with "Flat", for classes which use an underlying flat buffer for the dynamic memory of the array, or "Seg", for those that use a segmented buffer. Figure 7.1 illustrates the various layouts available.

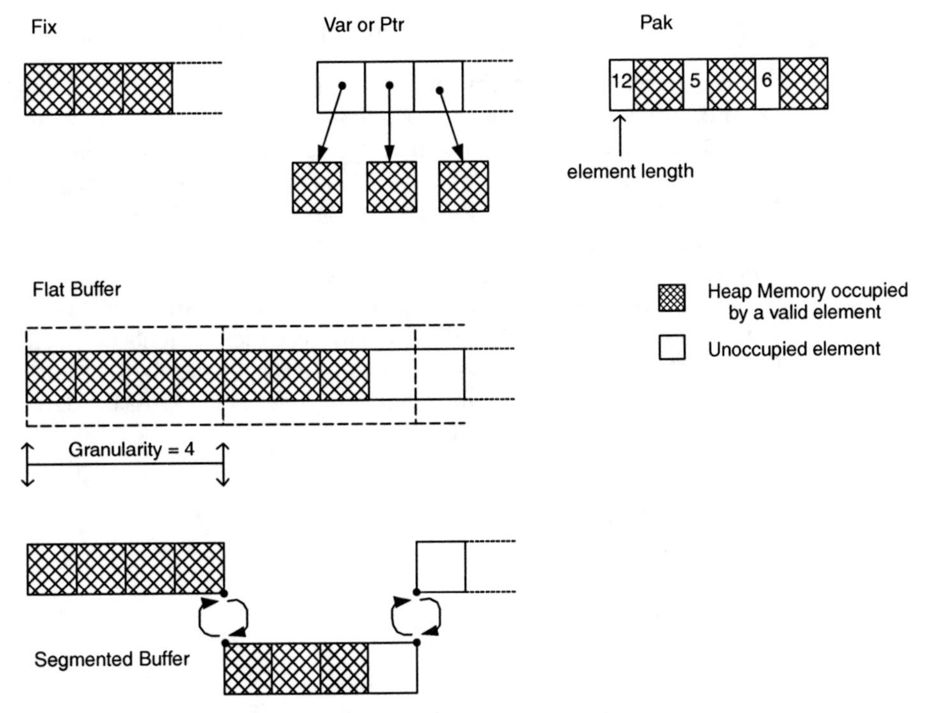

Figure 7.1 Memory layout of Symbian OS dynamic arrays

As I described above, the `RArray` classes are more efficient for simple arrays (flat arrays of fixed-length objects). For this reason, I will not discuss the `CArrayFixFlat` and `CArrayPtrFlat` classes at all, because you should use the `RArray` classes in preference.

However, there are other `CArrayX` classes which can be useful when you have variable-length elements or if you need to use a segmented buffer,[1] because there are no directly analogous `RArray` classes:

- `CArrayVarFlat` is used for variable-length elements referenced by pointer elements, using a flat memory layout for the array
- `CArrayVarSeg` is used for variable-length elements referenced by pointer elements, using a segmented array layout
- `CArrayPakFlat` is used for fixed- or variable-length elements that are stored in the flat array buffer itself, each element containing information about its length

[1] You may prefer to use a segmented buffer if reallocations are expected to be common, i.e. if the size of the array is likely to change frequently. If a single flat buffer is used, numerous reallocations may result in heap thrashing and copying. In addition, insertion and deletion in a segmented buffer can be more efficient than in a flat buffer, since it does not require all the elements after the modification point to be shuffled. However, you must weigh up the benefits of using a segmented memory buffer for the array against the other optimizations the flat `RArray` classes offer.

- `CArrayPtrSeg` is used for an array of pointers in a segmented array.

The inheritance hierarchy of the `CArrayX` classes is fairly straightforward. All of the classes are C classes and thus ultimately derive from `CBase`. Each class is a thin template specialization of one of the array base classes, `CArrayVarBase`, `CArrayPakBase` or `CArrayFixBase`. Thus, for example, `CArrayVarSeg<class T>` and `CArrayVarFlat<class T>` derive from `CArrayVar<class T>` which is a template specialization of `CArrayVarBase`, as shown in Figure 7.2.

`CArrayVarBase` owns an object that derives from `CBufBase`, the dynamic buffer base class, and is used to store the elements of the array. The object is a concrete instance of `CBufFlat` (a flat dynamic storage buffer) or `CBufSeg` (a segmented dynamic buffer). I'll discuss the dynamic buffer classes in more detail later.

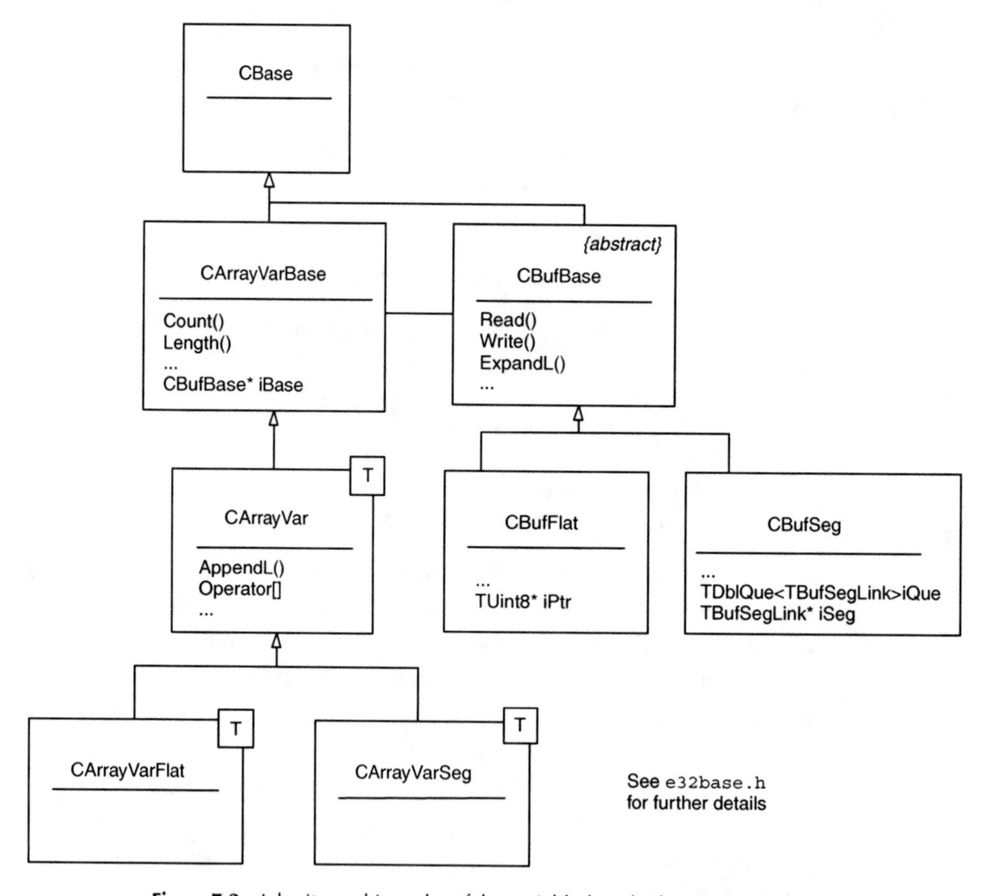

Figure 7.2 Inheritance hierarchy of the variable-length element array classes

Here is some example code showing how to manipulate the `CArrayPtrSeg` class. There's quite a lot of code but don't worry, it's quite straightforward, and I'll reuse it throughout the chapter to illustrate some of the other dynamic array classes. I've kept the sample code as brief as possible by omitting error checking and other code which isn't directly relevant to this chapter.

The example is of a very basic task manager which can be used to store tasks and execute them. The task manager class, `CTaskManager`, owns a dynamic array of pointers (in a `CArrayPtrSeg` object) to heap-based objects of the task class, `TTask`. The example shows the use of `AppendL()` and `InsertL()` to add an object to the array, `Delete()` to remove an element and `At()` and `operator[]` to access elements in the array.

The `TTask` class is rather empty, because I've implemented just the minimum amount of code for the example. You'll notice that it is a T class (as described in Chapter 1) and that I create objects on the heap and transfer ownership to the task manager array. Of course, I could have used any of the other `CArrayX` dynamic array classes to store the `TTask` objects, but I wanted to illustrate the use of the pointer array, particularly on cleanup. If the objects stored in the pointer array are not owned by another object, it is the responsibility of the array to destroy them when they are removed from the array or when the array itself is destroyed. You'll notice in the `CTaskManager::Delete()` method below that I store a pointer to the `TTask` object to be deleted, remove it from the array and then destroy it. In the destructor for `CTaskManager`, I call `ResetAndDestroy()` on the array, which empties the array, calling `delete` on every pointer.

```
class TTask // Basic T class, represents a task
    {
public:
    TTask(const TDesC& aTaskName);
    void ExecuteTaskL() {}; // Omitted for clarity
private:
    TBuf<10> iTaskName;
    };

TTask::TTask(const TDesC& aTaskName)
    { iTaskName.Copy(aTaskName); }

// Holds a dynamic array of TTask pointers
class CTaskManager : public CBase
    {
public:
    static CTaskManager* NewLC();
    ~CTaskManager();
public:
    void AddTaskL(TTask* aTask);
    void InsertTaskL(TTask* aTask, TInt aIndex);
    void RunAllTasksL();
    void DeleteTask(TInt aIndex);
public:
```

```
    inline TInt Count()
        {return (iTaskArray->Count());};
    inline TTask* GetTask(TInt aIndex)
        {return(iTaskArray->At(aIndex));};
private:
    void ConstructL();
    CTaskManager() {};
private:
    CArrayPtrSeg<TTask>* iTaskArray;
    };

const TInt KTaskArrayGranularity = 5;

CTaskManager* CTaskManager::NewLC()
    {
    CTaskManager* me = new (ELeave) CTaskManager();
    CleanupStack::PushL(me);
    me->ConstructL();
    return (me);
    }

CTaskManager::~CTaskManager()
    {// Cleanup the array
    if (iTaskArray)
        iTaskArray->ResetAndDestroy(); // Destroys objects through ptrs

    delete iTaskArray;
    }

void CTaskManager::ConstructL()
    {
    iTaskArray =
            new (ELeave) CArrayPtrSeg<TTask>(KTaskArrayGranularity);
    }

void CTaskManager::AddTaskL(TTask* aTask)
    { // Add a task to the end of array
      // No need to check that aTask! =NULL because CBufBase does this
      // before appending it
    iTaskArray->AppendL(aTask);
    }

void CTaskManager::InsertTaskL(TTask* aTask, TInt aIndex)
    { // Insert a task into a given element index
      // No assertion on aTask or aIndex because CArrayFixBase
      // and CBufBase do this
    iTaskArray->InsertL(aIndex, aTask);
    }

void CTaskManager::RunAllTasksL()
    { // Iterates all TTask objects and calls ExecuteTaskL()
    TInt taskCount = iTaskArray->Count();
    for (TInt index = 0; index < taskCount; index++)
        {
        (*iTaskArray)[index]->ExecuteTaskL();
        }
    }
```

```
void CTaskManager::DeleteTask(TInt aIndex)
   { // Removes the pointer from the array
     // The function stores a pointer to it so it can be destroyed
   TTask* task = iTaskArray->At(aIndex);
   if (task)
      {
      iTaskArray->Delete(aIndex); // Does not delete the object
      delete task;  // Deletes the object
      }
   }

// Calling code
void TestTaskManagerL()
   {
   CTaskManager* taskManager = CTaskManager::NewLC();

   // Add four tasks to the array
   _LIT(KTaskName, "TASKX%u");
   for (TInt index =0; index<4; index++)
      {
      TBuf<10> taskName;
      taskName.Format(KTaskName, index); // Names each task
      TTask* task = new (ELeave) TTask(taskName);
      CleanupStack::PushL(task);
      taskManager->AddTaskL(task);
      CleanupStack::Pop(task); // Now owned by the taskManager array
      }

   ASSERT(4==taskManager->Count()); // Chapter 16 discusses ASSERTs

   // Insert a task into element 3
   _LIT(KNewTask, "InsertedTask");
   TTask* insertedTask = new (ELeave) TTask(KNewTask);
   CleanupStack::PushL(insertedTask);
   taskManager->InsertTaskL(insertedTask, 3);
   CleanupStack::Pop(insertedTask); // Now owned by taskManager

   ASSERT(5==taskManager->Count());

   // Delete a task
   taskManager->DeleteTask(2);

   ASSERT(4==taskManager->Count());
   taskManager->RunAllTasksL();
   // Destroys the array (which itself destroys the TTasks it owns)
   CleanupStack::PopAndDestroy(taskManager);
   }
```

7.2 RArray<class T> and RPointerArray<class T>

RArray and RPointerArray are R classes, the characteristics of which were described in Chapter 1. The "R" of an R class indicates that it owns a resource, which in the case of these classes is the heap memory allocated to hold the array.

RArray objects themselves may be either stack- or heap-based. As with all R classes, when you have finished with an object you must call either the Close() or Reset() function to clean it up properly, that is, to free the memory allocated for the array. RArray::Close() frees the memory used to store the array and closes it, while RArray::Reset() frees the memory associated with the array and resets its internal state, thus allowing the array to be reused. It is acceptable just to call Reset() before allowing the array object to go out of scope, since all the heap memory associated with the object will have been cleaned up.

RArray<class T> comprises a simple array of elements of the same size. To hold the elements it uses a flat, vector-like block of heap memory, which is resized when necessary. This class is a thin template specialization of class RArrayBase.

RPointerArray<class T> is a thin template class deriving from RPointerArrayBase. It comprises a simple array of pointer elements which again uses flat, linear memory to hold the elements of the array, which should be pointers addressing objects stored on the heap. You must consider the ownership of these objects when you have finished with the array. If the objects are owned elsewhere, then it is sufficient to call Close() or Reset() to clean up the memory associated with the array. However, if the objects in the array are owned by it, they must be destroyed as part of the array cleanup, in a similar manner to the previous example for CArrayPtrSeg. This can be effected by calling ResetAndDestroy() which itself calls delete on each pointer element in the array.

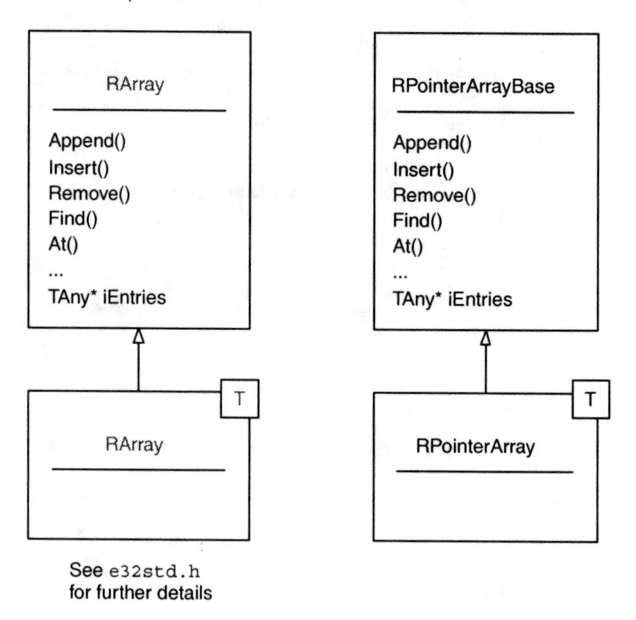

Figure 7.3 Inheritance hierarchy of RArray<T> and RPointerArray<T>

The RArray and RPointerArray classes are shown in Figure 7.3. They provide better searching and ordering than their CArrayX counterparts. The objects contained in RArray and RPointerArray may be ordered using a comparator function provided by the element class. That is, objects in the array supply an algorithm which is used to order them (typically implemented in a function of the element class) wrapped in a TLinearOrder<class T> package. It is also possible to perform lookup operations on the RArray and RPointerArray classes in a similar manner. The RArray classes have several Find() methods, one of which is overloaded to take an object of type TIdentityRelation<class T>. This object packages a function, usually provided by the element class, which determines whether two objects of type T match.

There are also two specialized classes defined specifically for arrays of 32-bit signed and unsigned integers, RArray<TInt> and RArray<TUint> respectively. These use the TEMPLATE_SPECIALIZATION macro to generate type-specific specializations over the generic RPointerArrayBase base class. The classes have a simplified interface, compared to the other thin template classes, which allows them to define insertion methods that do not need a TLinearOrder object and lookup methods that do not require a TIdentityRelation.

To illustrate how to use the RArray<class T> class, let's look at some example code. I've modified the CTaskManager class I used previously to use an RArray to store the TTask objects. First, here's the TTask class, which I've modified to add a priority value for the task, and functions for comparison and matching; where the code is identical to the previous example I've omitted it.

```
class TTask // Basic T class, represents a task
    {
public:
    TTask(const TDesC& aTaskName, const TInt aPriority);
    ...
public:
    static TInt ComparePriorities(const TTask& aTask1,
            const TTask& aTask2);
    static TBool Match(const TTask& aTask1, const TTask& aTask2);
private:
    TBuf<10> iTaskName;
    TInt iPriority;
    };

TTask::TTask(const TDesC& aTaskName, const TInt aPriority)
: iPriority(aPriority){iTaskName.Copy(aTaskName);}

// Returns 0 if both are equal priority,
// Returns a negative value if aTask1 > aTask2
// Returns a positive value if aTask1 < aTask2
TInt TTask::ComparePriorities(const TTask& aTask1, const TTask& aTask2)
    {
    if (aTask1.iPriority>aTask2.iPriority)
```

```
        return (-1);
    else if (aTask1.iPriority<aTask2.iPriority)
        return (1);
    else
        return (0);
    }

// Compares two tasks; returns ETrue if both iTaskName and
// iPriority are identical
TBool TTask::Match(const TTask& aTask1, const TTask& aTask2)
    {
    if ((aTask1.iPriority==aTask2.iPriority) &&
            (aTask1.iTaskName.Compare(aTask2.iTaskName)==0))
        {
        return (ETrue);
        }
    return (EFalse);
    }
```

Here's the task manager class, which has changed slightly because I'm now using `RArray<TTask>` rather than `CArrayPtrSeg<TTask>` to hold the set of tasks. Again, where the code is unchanged from the earlier example, I've omitted it for clarity.

```
class CTaskManager : public CBase
// Holds a dynamic array of TTask pointers
    {
public:
    static CTaskManager* NewLC();
    ~CTaskManager();
public:
    void AddTaskL(TTask& aTask);
    void InsertTaskL(TTask& aTask, TInt aIndex);
    void RunAllTasksL();
    void DeleteTask(TInt aIndex);
    void DeleteTask(const TTask& aTask);
public:
    inline TInt Count() {return (iTaskArray.Count());};
    inline TTask& GetTask(TInt aIndex) {return(iTaskArray[aIndex]);};
private:
    CTaskManager() {};
private:
    RArray<TTask> iTaskArray;
    };

CTaskManager::~CTaskManager()
    {// Close the array (free the memory associated with the entries)
    iTaskArray.Close();
    }

void CTaskManager::AddTaskL(TTask& aTask)
    {// Add a task to the end of array
    User::LeaveIfError(iTaskArray.Append(aTask));
    }

void CTaskManager::InsertTaskL(TTask& aTask, TInt aIndex)
```

```
    {// Insert a task in a given element
    User::LeaveIfError(iTaskArray.Insert(aTask, aIndex));
    }

void CTaskManager::RunAllTasksL()
    {// Sorts the tasks into priority order then iterates through them
     // and calls ExecuteTaskL() on each

    // Construct a temporary TLinearOrder object implicitly - the
    // equivalent of the following:
    // iTaskArray.Sort(TLinearOrder<TTask>(TTask::ComparePriorities));
    iTaskArray.Sort(TTask::ComparePriorities);

    TInt taskCount = iTaskArray.Count();
    for (TInt index = 0; index < taskCount; index++)
        {
        iTaskArray[index].ExecuteTaskL();
        }
    }

void CTaskManager::DeleteTask(const TTask& aTask)
    {// Removes all tasks identical to aTask from the array

    // Constructs a temporary TIdentityRelation object implicitly - the
    // equivalent of the following:
    // TInt foundIndex = iTaskArray.Find(aTask,
    //          TIdentityRelation<TTask>(TTask::Match));
    while (TInt foundIndex = iTaskArray.Find(aTask,
            TTask::Match)! =KErrNotFound)
        {
        iTaskArray.Remove(foundIndex);
        }
    }

void CTaskManager::DeleteTask(TInt aIndex)
    {// Removes the task at index aIndex from the array
    iTaskArray.Remove(aIndex);
    }
```

You'll notice the following changes:

- The calls to add and insert elements are within `User::LeaveIf-Error()` because those methods in `RArray` do not leave but instead return an error if a failure occurs (for example, if there is insufficient memory available to allocate the required space in the array).
- `CTaskManager::RunAllTasks()` sorts the array into descending priority order before iterating through the tasks and calling `Execute-TaskL()` on each.
- `CTaskManager` has an extra method that deletes all elements from the array where they are identical to the one specified: `Delete-Task(const TTask& aTask)`. This function uses the `Find()` function to match each element in the array against the task specified, deleting any it finds that are identical. Note that this function cannot leave.

Here's the modified version of code that uses the task manager class. It is quite similar to the previous code because the change to the encapsulated array class – which stores the tasks in `CTaskManager` – does not affect it:

```
void TestTaskManagerL()
    {
    CTaskManager* taskManager = CTaskManager::NewLC();

    // Add tasks to the array, use the index to set the task priority
    _LIT(KTaskName, "TASKX%u");
    for (TInt index=0; index<4; index++)
        {
        TBuf<10> taskName;
        taskName.Format(KTaskName, index);
        TTask task(taskName, index);
        taskManager->AddTaskL(task);
        }

    ASSERT(4==taskManager->Count());

    // Add a copy of the task at index 2
    // to demonstrate sorting and matching
    TBuf<10> taskName;
    taskName.Format(KTaskName, 2);
    TTask copyTask(taskName, 2);
    taskManager->AddTaskL(copyTask);

    ASSERT(5==taskManager->Count());

    taskManager->RunAllTasksL();

    // Remove both copies of the task
    taskManager->DeleteTask(copyTask);

    ASSERT(3==taskManager->Count());

    // Destroy the taskManager
    CleanupStack::PopAndDestroy(taskManager);
    }
```

7.3 Why Use `RArray` Instead of `CArrayX`?

The original `CArrayX` classes use `CBufBase`, which allows a varied dynamic memory layout for the array using either flat or segmented array buffers. However, `CBufBase` works with byte buffers and requires a `TPtr8` object to be constructed for every array access. This results in a performance overhead, even for a simple flat array containing fixed-length elements. Furthermore, for every method which accesses the array, there are a minimum of two assertions to check the parameters, even in release builds.

For example, to access a position in a CArrayFixX array, operator[] calls CArrayFixBase::At(), which uses an __ASSERT_ALWAYS statement to range-check the index and then calls CBufFlat::Ptr() which also asserts that the position specified lies within the array buffer. Likewise, adding an element to a CArrayFixFlat<class T> array using AppendL() runs through two assertion statements in CArrayFixBase::InsertL() and a further two __ASSERT_ALWAYS checks in CBufBase::InsertL(), which check that the appended object is valid and is being inserted within the range of the array.

Another issue is that a number of the array manipulation functions of CArrayX, such as AppendL(), can leave, for example when there is insufficient memory to resize the array buffer. While it is frequently acceptable to leave under these circumstances, in other cases (such as where the kernel uses the dynamic arrays) a leave is not allowed. In those circumstances, the leaving functions must be called in a TRAP macro to catch any leaves. As I described in Chapter 2, the TRAP macro has a performance overhead.

The RArray classes were added to Symbian OS to solve these issues for simple flat arrays. These classes have significantly better performance than CArrayX classes and do not need a TRAP harness. They are implemented as R classes for a lower overhead than C classes, because they do not need the characteristic features of a C class: zero-fill on allocation, a virtual function table pointer and creation on the heap. For reference, the Symbian OS class types (T, C, R and M) are discussed in more detail in Chapter 1.

The searching and ordering functions of the RArray classes were also optimized over those of the original classes and were made simpler to use.

Use RArray and RPointerArray instead of CArrayX except when you need support for segmented array memory and storage of elements of variable lengths (which is likely to be relatively rare).

7.4 Dynamic Descriptor Arrays

Descriptor arrays are dynamic arrays which are specialized for holding descriptors. These arrays extend the dynamic array classes and are defined in the Symbian OS Basic Application Framework Library (BAFL) component, which means that you must link against bafl.lib in order to use them. I'll describe them briefly – you'll find more documentation in your preferred SDK.

There are two types of descriptor array, both of which are provided for both 8-bit and 16-bit descriptors (the use of different width descriptors is discussed in more detail in Chapters 5 and 6):

- a pointer descriptor array
  This type of array holds only non-modifiable `TPtrC` descriptor elements. The pointer descriptors are added to the array, but the data they point to is not copied. The pointer descriptor array classes are `CPtrC8Array` and `CPtrC16Array` and derive from `CArrayFixFlat<TPtrC8>` and `CArrayFixFlat<TPtrC16>` respectively.
- a general descriptor array
  This type of array can hold any descriptor type, storing it as a non-modifiable element. That is, an `HBufC` copy is created for each descriptor added to the array; the array itself stores pointers to these heap descriptor copies. The abstract base class for a build-independent general descriptor array is `CDesCArray` (the explicit variants `CDesC16Array` and `CDesC8Array` may be used where necessary). These classes derive from `CArrayFixBase`, which was described earlier in this chapter. The concrete implementation classes are `CDesCXArrayFlat` (for flat array storage) or `CDesCXArraySeg` (for segmented storage), where `X` = 8, 16, or is not declared explicitly.

There are advantages and disadvantages of each type. General descriptor arrays are useful because you do not have to use a particular concrete descriptor type and thus can equally well store `HBufC`, `TPtrC` or `TBuf` objects in the array. These arrays take a copy of the original descriptor, which increases the amount of memory used compared to the pointer descriptor arrays, which do not take copies. However, it does mean that the original descriptor can then be safely discarded when using the general descriptor arrays. Pointer descriptor arrays do not take copies, so the descriptor elements must remain in memory for the lifetime of the array, otherwise it references invalid information.

7.5 Fixed-Length Arrays

Although this chapter is mostly about dynamic arrays, I'll briefly mention an alternative to them: fixed-length arrays. You may consider using fixed-length, C++ arrays when you know the number of elements that will occupy an array. Symbian OS provides the `TFixedArray` class, which wraps a fixed-length C++ array and adds range checking. Array access is automatically checked and can be performed in both release and debug builds. The checking uses assertion statements (as described in

Chapter 16) and a panic occurs if an attempt is made to use an out-of-range array index. The class can be used as a member of a CBase class (on the heap) or on the stack, since it is a T class.

The class is templated on the class type and the size of the array. Here's some example code to illustrate a fixed-size array containing five TTask objects. The array can be initialized by construction or by using the TFixedArray::Copy() function. The At() function performs range-checking in both release and debug builds, while operator[] checks for out-of-range indices in debug builds only.

```
class TTask
    {
public:
    TTask(TInt aPriority);
public:
    TInt iPriority;
    };

TTask::TTask(TInt aPriority)
: iPriority(aPriority){}

void TestFixedArray()
    {
    TTask tasks[5]={TTask(0), TTask(1), TTask(2), TTask(3), TTask(4)};
    // Wrap tasks with a range-checked TFixedArray
    TFixedArray<TTask, 5> taskArray(&tasks[0], 5);
    taskArray[1].iPriority = 3; // change priority
    taskArray[5].iPriority = 3;
    // Assertion fails -> panics debug builds (USER 133)
    taskArray.At(5).iPriority = 3;
    // Assertion fails -> panics debug & release builds
    }
```

These arrays are convenient where the number of elements is fixed and known at compile time. Once the array has been allocated, it cannot be resized dynamically, so insertion within the bounds of the array is guaranteed to succeed. Additionally, access to the array is fast in release mode.

Besides range-checking, the TFixedArray class has some useful additional functions which extend a generic C++ array. These include:

- Begin() and End() for navigating the array
- Count(), which returns the number of elements in the array
- Length(), which returns the size of an array element in bytes
- DeleteAll(), which invokes delete on each element of the array
- Reset(), which clears the array by filling each element with zeroes.

Besides having to know the size of the array in advance, the main drawbacks to the use of fixed-length arrays are that any addition to the array must occur at the end and that they do not support ordering and matching.

When working with fixed-length arrays, use `TFixedArray` instead of a generic C++ array to take advantage of bounds-checking and the utility functions provided.

7.6 Dynamic Buffers

Dynamic buffers are useful for storing binary data when its size is not fixed at compile time and it may need to expand to a potentially significant size at runtime. Descriptors or C++ arrays can be used to store binary data on Symbian OS, but these are not dynamically extensible; that is, a fixed-length C++ array cannot be expanded and a descriptor will panic if you attempt to write off the end of the array. You can use a heap descriptor, `HBufC`, and a modifiable pointer to write into it but even then you must manage the allocation of memory when you need to expand the array (as described in Chapter 5).

Dynamic buffers provide an alternative solution for binary data, but you should beware of using these classes as an alternative to descriptors for text data. The descriptor classes have been optimized for that purpose; in addition, dynamic buffer classes store data in 8-bit buffers, so you cannot use them comfortably for 16-bit Unicode strings.

When I described the underlying memory layout of the dynamic array classes, I mentioned that they use the `CBufBase`-derived classes `CBufFlat` and `CBufSeg`. `CBufBase` is an abstract class that provides a common interface to the dynamic buffers. This includes methods to insert, delete, read and write to the buffer. `CBufFlat` and `CBufSeg` are the concrete dynamic buffer classes. They are straightforward to use, as you'll see from the example code below. When instantiating an object using the static `NewL()` function, a granularity must be specified, which is the number of bytes by which the buffer will be reallocated when it needs to be resized. For a segmented buffer, the granularity determines the size of a segment.

Operations on dynamic buffers are all specified in terms of a buffer position, which is an integer value indicating a byte offset into the buffer data, and thus has a valid range from zero to the size of the buffer. The `InsertL()` and `DeleteL()` functions shuffle the data in the buffer after the insertion point. For this reason, any pointers to data in the dynamic buffers must be discarded when the data in the buffer is updated by insertion or deletion. As an example, consider an insertion into a flat buffer. This may potentially cause it to be reallocated, thus invalidating any pointers to data in the original heap cell. Likewise, deletion of buffer data causes data after the deletion point to move up the buffer. For this reason, it is sensible to reference data in the dynamic buffers only in terms of the buffer position.

Let's take a look at some example code for the dynamic buffers which stores 8-bit data received from a source of random data. The details of the example are not that important; in fact it's somewhat contrived, and its main purpose is to illustrate the use of the `InsertL()`, `Delete()`, `Compress()`, `ExpandL()`, `Read()`, and `Write()` functions.

```
// Returns random data in an 8-bit heap descriptor of length = aLength
HBufC8* GetRandomDataLC(TInt aLength); // Defined elsewhere

void PrintBufferL(CBufBase* aBuffer)
    {
    aBuffer->Compress();
    // Compress to free unused memory at the end of a segment
    TInt length = aBuffer->Size();
    HBufC8* readBuf = HBufC8::NewL(length);
    TPtr8 writable(readBuf->Des());
    aBuffer->Read(0, writable);
    ... // Omitted. Print to the console
    delete readBuf;
    }

    void TestBuffersL()
    {
    __UHEAP_MARK; // Heap checking macro to test for memory leaks

    CBufBase* buffer = CBufSeg::NewL(16); // Granularity = 16
    CleanupStack::PushL(buffer);       // There is no NewLC() function

    HBufC8* data = GetRandomDataLC(32);// Data is on the cleanup stack

    buffer->InsertL(0, *data);
    // Destroy original. A copy is now stored in the buffer
    CleanupStack::PopAndDestroy(data);

    PrintBufferL(buffer);

    buffer->ExpandL(0, 100); // Pre-expand the buffer

    TInt pos = 0;
    for (TInt index = 0; index <4; index++, pos+16)
        {// Write the data in several chunks
        data = GetRandomDataLC(16);
        buffer->Write(pos, *data);
        CleanupStack::PopAndDestroy(data); // Copied so destroy here
        }

    PrintBufferL(buffer);
    CleanupStack::PopAndDestroy(buffer);    // Clean up the buffer
    __UHEAP_MARKEND;                        // End of heap checking
    }
```

The dynamic buffer, of type CBufSeg, is instantiated at the beginning of the TestBuffersL() function by a call to CBufSeg::NewL(). A 32-byte block of random data is retrieved from GetRandomDataLC() and inserted into the buffer at position 0, using InsertL().

The example also illustrates how to pre-expand the buffer using ExpandL() to allocate extra space in the buffer at the position specified (alternatively, you can use ResizeL(), which adds extra memory at the end). These methods are useful if you know there will be a number of insertions into the buffer and you wish to make them atomic. ExpandL() and ResizeL() perform a single allocation, which may fail, but if the buffer is expanded successfully then data can be added to the array using Write(), which cannot fail. This is useful to improve performance, since making a single call which may leave (and thus may need to be called in a TRAP) is far more efficient than making a number of InsertL() calls, each of which needs a TRAP.[2] Following the buffer expansion, random data is retrieved and written to the buffer in a series of short blocks.

The PrintBufferL() function illustrates the use of the Compress(), Size() and Read() methods on the dynamic buffers. The Compress() method compresses the buffer to occupy minimal space, freeing any unused memory at the end of a segment for a CBufSeg, or the end of the flat contiguous buffer for CBufFlat. It's a useful method for freeing up space when memory is low or if the buffer has reached its final size and cannot be expanded again. The example code uses it in PrintBufferL() before calling Size() on the buffer (and using the returned value to allocate a buffer of the appropriate size into which data from Read() is stored). The Size() method returns the number of heap bytes allocated to the buffer, which may be greater than the actual size of the data contained therein, because the contents of the buffer may not fill the total allocated size. The call to Compress() before Size() thus retrieves the size of the data contained in the buffer rather than the entire memory size allocated to it.

To retrieve data from the buffer, you can use the Ptr() function, which returns a TPtr8 for the given buffer position up to the end of the memory allocated. For a CBufFlat this returns a TPtr8 to the rest of the buffer, but for CBufSeg it returns only the data from the given position to the end of that segment. To retrieve all the data in a segmented buffer using Ptr(), you must use a loop which iterates over every allocated segment. Alternatively, as I've done in PrintBufferL(), the Read() method transfers data from the buffer into a descriptor, up to the length of the descriptor or the maximum length of the buffer, whichever is smaller.

---

[2] In addition, for CBufFlat, multiple calls to InsertL() will fragment the heap whereas a single ExpandL() or ResizeL() call will allocate all the memory required in a single contiguous block.

7.7 Summary

This chapter discussed the use of the Symbian OS dynamic array classes, which allow collections of data to be manipulated, expanding as necessary as elements are added to them. Unlike C++ arrays, dynamic arrays do not need to be created with a fixed size. However, if you do know the size of a collection, Symbian OS provides the `TFixedArray` class to represent a fixed-length array which extends simple C++ arrays to provide bounds-checking.

The chapter described the characteristics and use of the dynamic container classes `RArray` and `RPointerArray` and also discussed the `CArrayX` classes which the `RArray` classes supersede. The `RArray` classes were introduced to Symbian OS for enhanced performance over the `CArrayX` classes. They have a lower overhead because they do not construct a `TPtr8` for each array access, have fewer assertion checks and no leaving methods and are implemented as R classes, which tend to have a lower overhead than C classes. The `RArray` classes also have improved search and sort functionality and should be preferred over `CArrayX` classes. However, the `CArrayX` classes may still be useful when dealing with variable-length elements or segmented memory, because there are no `RArray` analogues.

The chapter also discussed the descriptor array classes, `CPtrC8Array` and `CPtrC16Array`, and the dynamic buffers, `CBufFlat` and `CBufSeg`.

All the dynamic array and buffer classes in Symbian OS are based on the thin template idiom (see Chapter 19). The use of lightweight templates allows the elements of the dynamic arrays to be objects of any type, such as pointers to `CBase`-derived objects, or T and R class objects.

8

Event-Driven Multitasking Using Active Objects

Light is the task where many share the toil

Homer

Active objects are a fundamental part of Symbian OS. This chapter explains why they are so important, and how they are designed for responsive and efficient event handling. Active objects are intended to make life easy for application programmers, and this chapter alone gives you sufficient knowledge to work with them within an application framework or derive your own simple active object class. Chapter 9 will be of interest if you want to write or work with more complex active objects: it discusses the responsibilities of active objects, asynchronous service providers and the active scheduler in detail, and reviews the best strategies for long-running or low-priority tasks. System-level programmers wishing to understand the Symbian OS client–server architecture and lower-level system design should read both this chapter and the following one.

8.1 Multitasking Basics

First of all, what are active objects for? Well, let's go back to basics. Consider what happens when program code makes a function call to request a service. The service can be performed either synchronously or asynchronously. When a synchronous function is called, it performs a service to completion and returns directly to its caller, usually returning an indication of its success or failure (or leaving, as discussed in Chapter 2). An asynchronous function submits a request as part of the function call and returns to its caller – but completion of that request occurs some time later. Before the request completes, the caller may perform other processing or it may simply wait, which is often referred to as "blocking". Upon completion, the caller receives a signal which indicates the success or failure of the request. This signal is known as an event, and the code can be said to be event-driven. Symbian OS, like other operating systems, uses event-driven code extensively both at a high level, e.g. for user interaction, and at a lower, system level, e.g. for asynchronous communications input and output.

Before considering active objects further, let's consider how code actually "runs". A thread is a fundamental unit of execution, which runs within a process. A process has its own address space and may have one or more threads independently executing code within it. When a process is created, a single primary thread is initialized within it. Other threads may then be created, as described in Chapter 10. Code executing in that process accesses virtual memory addresses which are mapped for that process to physical locations in hardware by the memory management unit. The writable memory of one process is not normally accessible to another process, thus "protecting" processes from each other. However, multiple threads running in the same process are not isolated from each other in the same way because they share the memory mapped for the process in which they run. This means that they can access each other's data, which is useful, but they can also accidentally scribble on it, which is not.

On Symbian OS, threads are scheduled pre-emptively by the kernel, which runs the highest priority thread eligible. Each thread may be suspended while waiting for a given event to occur and may resume whenever appropriate. The kernel controls thread scheduling, allowing the threads to share the system resources by time-slice division, pre-empting the running of a thread if another, higher priority thread becomes eligible to run. This constant switching of the running thread is the basis of pre-emptive multitasking, which allows multiple servers and applications to run simultaneously. A context switch occurs when the currently running thread is suspended (for example, if it is blocked, has reached the end of its time-slice, or a higher priority thread becomes ready to run) and another thread is made current by the scheduler. The context switch incurs a runtime overhead in terms of the kernel scheduler and, potentially, the memory management unit and hardware caches, if the original and replacing threads are executing in different processes.

8.2 Event-Driven Multitasking

Moving up a level, let's look at some typical examples of events and event-driven multitasking. Events can come from external sources, such as user input or hardware peripherals that receive incoming data. They can also be generated by software, for example by timers or completed asynchronous requests. Events are managed by an event handler, which, as its name suggests, waits for an event and then handles it.

An example of an event handler is a web browser application, which waits for user input and responds by submitting requests to receive web pages which it then displays. The web browser may use a system server, which waits to receive requests from its clients, services them and returns to waiting for another request. The system server submits requests, e.g. I/O requests, to other servers, which later generate completion events. Each of the software components described is event-driven. They need to be responsive to user input and responsive to requests from the system (for example, from the communications infrastructure).

In response to an event, the event handler may request another service. This service will later cause another event, or may indicate that the service has completed, which may cause another event in a different part of the system. The operating system must have an efficient event-handling model to handle each event as soon as possible after it occurs and, if more than one event occurs, in the most appropriate order. It is particularly important that user-driven events are handled rapidly to give feedback and a good user experience. Between events, the system should wait in a low power state. This avoids polling constantly, which can lead to significant power drain and should be avoided on a battery-powered device. Instead the software should allow the operating system to move to an idle mode, while it waits for the next event.

On hardware running Symbian OS, resources are more limited than on a typical desktop PC. Thus, on Symbian OS, besides the requirements to be responsive and handle power consumption carefully, it is also important that the memory used by event-handling code is minimized and that processor resources are used efficiently. Active objects assist with efficient programming by providing a model for lightweight, event-driven multitasking.

Active objects encapsulate the traditional wait loop inside a class. They were designed such that a switch between active objects that run in the same thread incurs a lower overhead than a thread context switch.[1] This makes active objects preferable for event-driven multitasking on Symbian OS.

Apart from the runtime expense of a context switch, using pre-emptive multithreading for event handling can be inconvenient because of the need to protect shared objects with synchronization primitives such as mutexes or semaphores. Additionally, resource ownership is thread-relative by default on Symbian OS. If a file is opened by the main thread it will not be possible for a different thread in the process to use it without the handle being explicitly shared through a call to `RSessionBase::Share()` (and some Symbian OS servers do not support session sharing at all). Because of this restriction, it may be

[1] The difference in speed between a context switch between threads and transfer of control between active objects in the same thread can be of a factor of 10 in favor of active objects. In addition, the space overhead for a thread can be around 4 KB kernel-side and 8 KB user-side for the program stack, while the size of an active object may be only a few hundred bytes, or less.

very difficult to use multiple threads as a viable method for event-driven multitasking. More detail on threads can be found in Chapter 10, while the client–server model is discussed in Chapters 11 and 12.

On Symbian OS, active objects multitask cooperatively and, consequently, there is no need for synchronization protection of shared resources. In addition, because active objects run in the same thread, memory and objects may be shared more readily. Active objects still run independently of each other, despite existing in the same thread, in much the same way as threads are independent of each other in a process.

On Symbian OS, the use of active objects for event-handling multitasking is ideal because they are designed for efficiency and, if used correctly, to be responsive. In general, a Symbian OS application or server will consist of a single main event-handling thread. A set of active objects run in the thread, each representing a task. Each active object requests an asynchronous service, waits while it is serviced, handles the request completion event and communicates with other tasks as necessary.

Some events require a response within a guaranteed time, regardless of any other activity in the system. This is called "real-time" event-handling. For example, a real-time task may be required to keep the buffer of a sound driver supplied with sound data – a delay in response delays the sound decoding, which results in it breaking up. Other typical real-time requirements may be even more strict, say for low-level telephony. These tasks have, in effect, different requirements for real-time responses, which can be represented by task priorities. Higher-priority tasks must always be able to pre-empt lower-priority tasks in order to guarantee to meet their real-time requirements. The shorter the response time required, the higher the priority that should be assigned to a task.

However, once an active object is handling an event, it may not be pre-empted by the event handler of another active object[2], which means that they are not suitable for real-time tasks. On Symbian OS, real-time tasks should be implemented using high-priority threads and processes, with the priorities chosen as appropriate for relative real-time requirements.

Active objects are used on Symbian OS to simplify asynchronous programming and make it easy for you to write code to submit asynchronous requests, manage their completion events and process the result. They are well suited for lightweight event-driven programming, except where a real-time, guaranteed response is required.

[2] Note that, although the active objects within a thread run cooperatively without pre-emption, on Symbian OS the thread in which they run is scheduled pre-emptively.

8.3 Working with Active Objects

Let's move on now to consider in more detail how active objects work, and how to use them. A typical Symbian OS application or server consists of a single event-handling thread running a scheduler (the "active scheduler") which coordinates one or more active objects. Each active object requests an asynchronous service and handles the resulting completion event some time after the request. It also provides a way to cancel an outstanding request and may provide error handling for exceptional conditions.

An active object class must derive from class CActive, which is defined in e32base.h (shown here with only the relevant methods, for clarity). The next chapter discusses details of the base class further.

```
class CActive : public CBase
    {
public:
    enum TPriority
        {
        EPriorityIdle=-100,
        EPriorityLow=-20,
        EPriorityStandard=0,
        EPriorityUserInput=10,
        EPriorityHigh=20,
        };
public:
    IMPORT_C ~CActive();
    IMPORT_C void Cancel();
    ...
    IMPORT_C void SetPriority(TInt aPriority);
    inline TBool IsActive() const;
    ...
protected:
    IMPORT_C CActive(TInt aPriority);
    IMPORT_C void SetActive();
    virtual void DoCancel() =0;
    virtual void RunL() =0;
    IMPORT_C virtual TInt RunError(TInt aError);
public:
    TRequestStatus iStatus;
    ...
    };
```

Construction

Like threads, active objects have a priority value to determine how they are scheduled. Classes deriving from CActive must call the protected constructor of the base class, passing in a parameter to set the priority of the active object.

When the asynchronous service associated with the active object completes, it generates an event. The active scheduler detects events, determines which active object is associated with each event, and calls the appropriate active object to handle the event. I'll describe this in more detail in the later section on event handling. While an active object is handling an event, it cannot be pre-empted[3] until the event handler function has returned back to the active scheduler.

It is quite possible that a number of events may complete before control returns to the scheduler. The scheduler must resolve which active object gets to run next; it does this by ordering the active objects using their priority values. If multiple events have occurred before control returns to the scheduler, they are handled sequentially in order of the highest active object priority, rather than in order of completion. Otherwise, an event of low priority that completed just before a more important one would supplant the higher-priority event for an undefined period, depending on how much code it executed to run to completion.

A set of priority values are defined in the `TPriority` enumeration of class `CActive`. For the purposes of this chapter, and in general, you should use the priority value `EPriorityStandard` (=0) unless you have good reason to do otherwise. The next chapter discusses the factors you should consider when setting the priority of your active objects on construction or by making additional calls to `CActive::SetPriority()`.

As part of construction, the active object code should call a static function on the active scheduler, `CActiveScheduler::Add()`. This will add the object to a list maintained by the active scheduler of event-handling active objects on that thread.

An active object typically owns an object to which it issues requests that complete asynchronously, generating an event, such as a timer object of type `RTimer`. This object is generally known as an asynchronous service provider and it may need to be initialized as part of construction. Of course, if the initialization can fail, you should perform it as part of the second-phase construction, as described in Chapter 4.

Submitting Requests

An active object class supplies public methods for callers to initiate requests. These will submit requests to the asynchronous service provider associated with the active object, using a well-established pattern, as follows:

1. Request methods should check that there is no request already submitted before attempting to submit another. Each active object

[3] While this is true under most circumstances, it is possible to nest a separate active scheduler within the event handler of an active object, and use the nested active scheduler to receive other active objects' events. This is used in Uikon for modal "waiting" dialogs, as discussed further in the next chapter. However, this technique can cause complications and should be used with caution. For the purposes of discussion in this chapter, you should simply consider that all active object event handling is non-preemptive.

can only ever have one outstanding request, for reasons I'll discuss further in the next chapter. Depending on the implementation, the code may:

- panic if a request has already been issued, if this scenario could only occur because of a programming error
- refuse to submit another request (if it is legitimate to attempt to make more than one request)
- cancel the outstanding request and submit the new one.

2. The active object should then issue the request to the service provider, passing in its `iStatus` member variable as the `TRequestStatus&` parameter.
3. If the request is submitted successfully, the request method then calls the `SetActive()` method of the `CActive` base class, to indicate to the active scheduler that a request has been submitted and is currently outstanding. This call is not made until after the request has been submitted.

Event Handling

Each active object class must implement the pure virtual `RunL()` member method of the `CActive` base class. When a completion event occurs from the associated asynchronous service provider and the active scheduler selects the active object to handle the event, it calls `RunL()` on the active object.[4] The function has a slightly misleading name because the asynchronous function has already run. Perhaps a clearer description would be `HandleEventL()` or `HandleCompletionL()`.

Typical implementations of `RunL()` determine whether the asynchronous request succeeded by inspecting the completion code in the `TRequestStatus` object (`iStatus`) of the active object, a 32-bit integer value. Depending on the result, `RunL()` usually either issues another request or notifies other objects in the system of the event's completion; however, the degree of complexity of code in `RunL()` can vary considerably. Whatever it does, once `RunL()` is executing, it cannot be pre-empted by other active objects' event handlers. For this reason, the code should complete as quickly as possible so that other events can be handled without delay.

Figure 8.1 illustrates the basic sequence of actions performed when an active object submits a request to an asynchronous service provider that later completes, generating an event which is handled by `RunL()`.

---

[4] Advanced Windows programmers will recognize the pattern of a message loop and message dispatch which drives a Win32 application. On Symbian OS, the active scheduler takes the place of the Windows message loop and the `RunL()` function of an active object acts as a message handler.

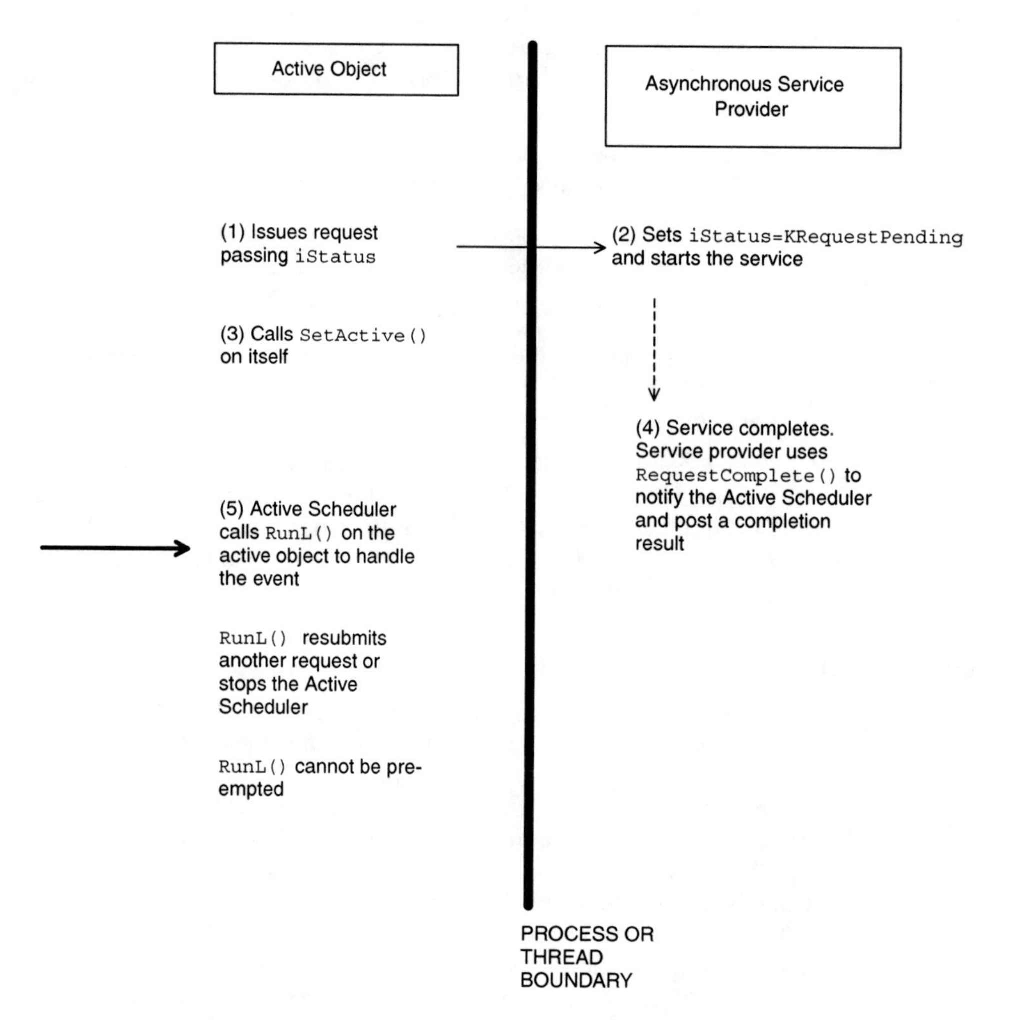

Figure 8.1 Process of submitting a request to an asynchronous service provider, which generates an event on completion

Each active object class must implement the pure virtual `RunL()` member method of the `CActive` base class to handle completion events.

Cancellation

The active object must be able to cancel any outstanding asynchronous requests it has issued, for example, if the application thread in which it is running is about to terminate. The `CActive` base class implements a `Cancel()` method which calls the pure virtual `DoCancel()` method (which the derived active object class must implement) and waits for the request's early completion. Any implementation of `DoCancel()` should call the appropriate cancellation method on the asynchronous service provider. `DoCancel()` can also include other processing, but should not leave or allocate resources and should not carry out any lengthy operations. It's a good rule to restrict the method to cancellation and any necessary cleanup associated with the request, rather than implementing any sophisticated functionality. This is because a destructor should call `Cancel()`, as described in later, and may have already cleaned up resources that `DoCancel()` may require.

It isn't necessary to check whether a request is outstanding for the active object before calling `Cancel()`, because it is safe to do so even if it isn't currently active. The next chapter discusses in detail the semantics of canceling active objects.

Error Handling

From Symbian OS v6.0 onwards, the `CActive` class provides a virtual `RunError()` method which the active scheduler calls if a leave occurs in the `RunL()` method of the active object. The method takes the leave code as a parameter and returns an error code to indicate whether the leave has been handled. The default implementation does not handle the leave and simply returns the leave code passed to it. If the active object can handle any leaves occurring in `RunL()`, it should do so by overriding the default implementation and returning `KErrNone` to indicate that the leave has been handled.

If `RunError()` returns a value other than `KErrNone`, indicating that the leave has yet to be dealt with, the active scheduler calls its `Error()` function to handle it. The active scheduler does not have any contextual information about the active object with which to perform error handling. For this reason, it is generally preferable to manage error recovery within the `RunError()` method of the associated active object, where more context is usually available.

Destruction

The destructor of a `CActive`-derived class should always call `Cancel()` to terminate any outstanding requests as part of cleanup code. This should ideally be done before any other resources owned by the active object are destroyed, in case they are used by the service provider or the `DoCancel()` method. The destructor code should, as usual, free all resources owned by the object, including any handle to the asynchronous service provider.

The `CActive` base class destructor is virtual and its implementation checks that the active object is not currently active. It panics if any request is outstanding, that is, if `Cancel()` has not been called. This catches any programming errors which could lead to the situation where a request completes after the active object to handle it has been destroyed. This would otherwise result in a "stray signal", described further in Chapter 9, where the active scheduler cannot locate an active object to handle the event.

Having verified that the active object has no issued requests outstanding, the `CActive` destructor removes the active object from the active scheduler.

The destructor of a `CActive`-derived class should always call `Cancel()` to terminate any outstanding requests before cleanup proceeds.

8.4 Example Code

The example below illustrates the use of an active object class to wrap an asynchronous service, in this case a timer provided by the `RTimer` service. In fact, Symbian OS already supplies an abstract active object class, `CTimer`, which wraps `RTimer` and can be derived from, specifying the action required when the timer expires. However, I've created a new class, `CExampleTimer`, because it's a straightforward way of illustrating active objects. Figure 8.2 illustrates the classes involved and their relationship with the active scheduler.

When the timer expires, the `RunL()` event handler checks the active object's `iStatus` result and leaves if it contains a value other than `KErrNone` so the `RunError()` method can handle the problem. In this case, the error handling is very simple: the error returned from the request is logged to file. This could have been performed in the `RunL()` method, but I've separated it into `RunError()` to illustrate how to use the active object framework to split error handling from the main logic of the event handler. If no error occurred, the `RunL()` event handler logs the timer completion to debug output using `RDebug::Print()` (described in Chapter 17) and resubmits the timer request with the stored interval value. In effect, once the timer request has started, it continues to expire and be resubmitted until it is stopped by a call to the `Cancel()` method on the active object.

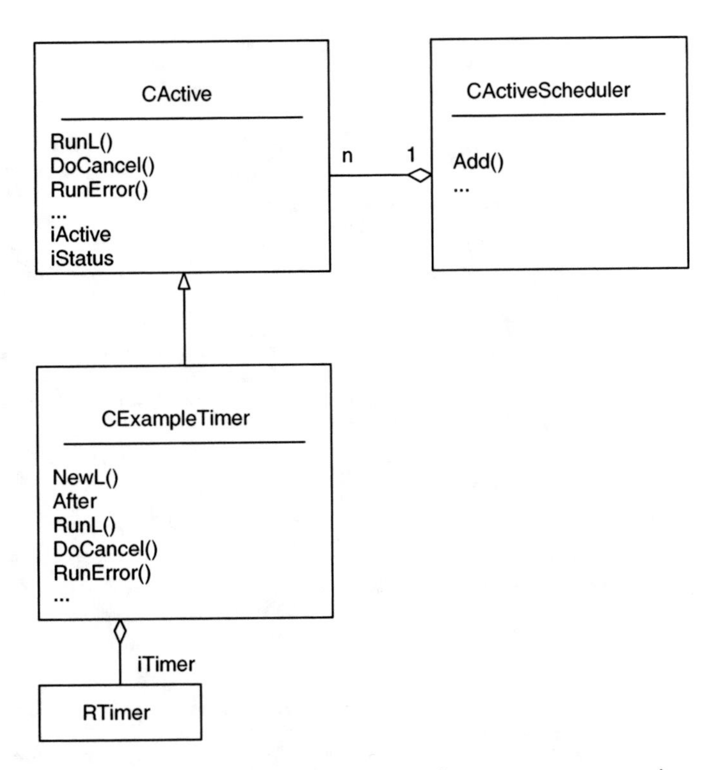

Figure 8.2 `CExampleTimer` and its relationship with `RTimer`, `CActive` and `CActiveScheduler`

```
_LIT(KExampleTimerPanic, "CExampleTimer");

class CExampleTimer : public CActive
    {
public:
    ~CExampleTimer();
    static CExampleTimer* NewL();
    void After(TTimeIntervalMicroSeconds32& aInterval);
protected:
    CExampleTimer();
    void ConstructL();
protected:
    virtual void RunL();      // Inherited from CActive
    virtual void DoCancel();
    virtual TInt RunError(TInt aError);
private:
    RTimer iTimer;
    TTimeIntervalMicroSeconds32 iInterval;
    };

CExampleTimer::CExampleTimer()
: CActive(EPriorityStandard) { CActiveScheduler::Add(this); }

void CExampleTimer::ConstructL()
```

```
    {// Create the asynchronous service provider
    User::LeaveIfError(iTimer.CreateLocal());
    }
CExampleTimer* CExampleTimer::NewL()
    {...} // Code omitted for clarity

CExampleTimer::~CExampleTimer()
    {
    Cancel();
    iTimer.Close();
    }

void CExampleTimer::After(TTimeIntervalMicroSeconds32& aInterval)
    {// Only allow one timer request to be submitted at a time
     // Caller must call Cancel() before submitting another
    __ASSERT_ALWAYS(!IsActive(), User::Panic(KExampleTimerPanic,
            KErrInUse));
    iInterval = aInterval;
    iTimer.After(iStatus,aInterval); // Set the RTimer
    SetActive(); // Mark this object active
    }

void CExampleTimer::RunL()
    {// If an error occurred (admittedly unlikely)
     // deal with the problem in RunError()
    User::LeaveIfError(iStatus.Int());

    // Otherwise, log the timer completion and resubmit the timer
    _LIT(KTimerExpired, "Timer Expired\n");
    RDebug::Print(KTimerExpired); // See Chapter 17 for more details
    iTimer.After(iStatus, iInterval);
    SetActive();
    }

void CExampleTimer::DoCancel()
    {// Cancel the timer
    iTimer.Cancel();
    }

TInt CExampleTimer::RunError(TInt aError)
    {// Called if RunL() leaves, aError contains the leave code
    _LIT(KErrorLog, "Timer error %d");
    RDebug::Print(KErrorLog, aError); // Logs the error
    return (KErrNone);                 // Error has been handled
    }
```

The example is simplistic but it demonstrates the use of an active object to wrap an asynchronous function, in this case, timer completion. Here's how a client of class `CExampleTimer` can expect to use the active object, which it stores as a member variable called `iExampleTimer`, illustrating the transparency of active object classes from a client's perspective:

```
// Class CClient has a member variable CExampleTimer* iExampleTimer
void CClient::StartRepeatingTimerL()
    {
    iExampleTimer = CExampleTimer::NewL();
```

```
    iExampleTimer->After(1000000);
    }

void CClient::StopRepeatingTimer()
    {
    iExampleTimer->Cancel();
    delete iExampleTimer;
    iExampleTimer = NULL; // Prevents re-use or double deletion
    }
```

8.5 Threads Without an Active Scheduler

Most threads running on Symbian OS have an active scheduler, which is usually created implicitly by a framework (e.g. CONE for the GUI framework). However, if you are implementing a server, you have to create and start an active scheduler explicitly before you can use active objects. Likewise, console-based test code may not use active objects directly itself, but must create an active scheduler in its main thread if it depends on components which do use active objects.

There are a few threads in Symbian OS which intentionally do not have an active scheduler and thus cannot use active objects or components that use them:

- The Java implementation does not support an active scheduler and native Java methods may not use active objects. It is permissible to make calls in code to C++ servers which do use them, since these run in a separate thread which has a supporting active scheduler.
- The C Standard Library, STDLIB, thread has no active scheduler and thus standard library code cannot use active objects. Functions provided by the Standard Library may however be used in active object code, for example in an initialization or a `RunL()` method. The functions should be synchronous and return quickly, as required by all active object implementations.
- OPL does not provide an active scheduler and C++ extensions to OPL (OPXs) must not use active objects or any component which uses them.

8.6 Application Code and Active Objects

The active object model is very easy to use without needing a full understanding of how it works. In this chapter, I've described how to handle events resulting from the completion of asynchronous functions on Symbian OS. This involves defining `CActive`-derived classes, and providing the event handler and additional code for construction, destruction, error-handling and cancellation. However, application code requires some understanding of active objects and requires implementation of active object code that is incidental to the purpose of the application. In addition, many other operating systems do not encapsulate event handling in the same way. This means that Symbian OS code is not directly portable to other operating systems.

For this reason, Symbian OS provides frameworks to hide active object code from high-level application code by defining handler interfaces (often mixin classes as described in Chapter 1) which a client implements. The frameworks hide the active objects, implementing initialization, cleanup and error handling, and performing generic event processing in the `RunL()` method. In `RunL()` the framework calls methods on the client-implemented interface to respond to the completed event. The appropriate client object receives the call, having "registered" with the framework by passing a reference or pointer to its implementation of the interface. The one rule the client should be aware of is that the interface methods should be implemented to complete quickly so that the framework can handle other events without delay.

An example of this may be seen when application code runs within the GUI framework (CONE), which uses active objects to handle events associated with user input and system requests such as window redraw requests, originating from the window server. The framework implements the `RunL()` method to process each event, and calls an appropriate virtual function such as `OfferKeyEventL()` or `HandlePointerEventL()` in the associated control. As an application programmer, all you have to do is implement the functions defined by the framework and it takes care of the mechanics of event handling.

Likewise, server code runs inside the Symbian OS system server framework and handles requests from client messages. Server code often implements its own additional set of internal active objects, to make requests to asynchronous service providers while still remaining responsive to incoming client requests. The client–server architecture is discussed in detail in Chapters 11 and 12.

8.7 Summary

Symbian OS makes it very easy to write event-driven code without getting too involved in the implementation of the active objects which handle the events. This chapter set out to explain the basics of how active objects work, why you should prefer them to threads for multitasking and how to implement a basic active object class. If you are likely to write server code, implement an asynchronous service provider or perform more advanced GUI programming, you will probably be interested in the next chapter, which goes deeper into the responsibilities and roles of the main components – active objects, asynchronous service providers and the active scheduler.

This chapter explained that:

- Active objects were designed for lightweight, responsive, power-efficient event-handling and are used by most applications and servers, generally in a single thread.
- Active objects are scheduled cooperatively in a single thread and, in consequence, are easy to program because there is no need to use synchronization primitives to prevent concurrent access to shared resources.
- The kernel schedules threads pre-emptively but these are not generally used for event-driven multitasking.
- The active object framework on Symbian OS has a modular design that decouples event completion processing, performed by the active scheduler, from individual event handling, performed by active objects.
- The constructor of the active object class should set its priority and add it to the active scheduler. Any initialization required by the asynchronous service provider should be performed in a second-phase constructor.
- Active objects encapsulate an asynchronous service provider and the event handling necessary when a request to that service provider completes.
- Active objects provide request initiation functions. These usually conform to a pattern: performing a check for no previous outstanding requests, submitting the request with its `iStatus` member variable as a parameter, and setting the state of the active object to indicate that a request has been issued.
- Event handling is performed by implementing the pure virtual `RunL()` method defined in the `CActive` base class; `RunL()` is called by the active scheduler some time after the request completes.
- Because active objects cannot pre-empt each other, `RunL()` should complete as quickly as possible so that other active object events can be handled without delay.
- The pure virtual `DoCancel()` method must also be implemented by deriving classes and should call an appropriate method on the asynchronous service provider to cancel the request.

- The default "do nothing" implementation of the virtual `RunError()` method on the base class should be overridden if the derived class can reasonably handle leaves which occur within its `RunL()` function.
- Symbian OS uses framework code to conceal the active object idiom. For example, the GUI framework dispatches incoming events to be handled by functions implemented by an associated control, which may be extended by individual applications as required.

The next chapter looks "under the hood" at active objects, discusses the responsibilities of active objects, asynchronous service providers and the active scheduler, and illustrates best practice when writing and using active objects. Chapter 10 has more information about Symbian OS threads and processes, while Chapters 11 and 12 build on an understanding of active objects, threads and processes, to describe the Symbian OS client–server architecture in detail.

9

Active Objects under the Hood

Do not hack me as you did my Lord Russell

The last words of the Duke of Monmouth (1649–1685) addressed to his executioner

The previous chapter introduced active objects and described the basics, such as how to use them, how to derive a simple active object class and why active objects are used by Symbian OS as a lightweight alternative to threads for event-driven multitasking. This chapter considers active objects in more detail and discusses some commonly used active object idioms.

First of all, let's examine in detail the responsibilities of active objects, asynchronous service providers and the active scheduler and walk through how they fit together. The previous chapter made the following main points:

- Symbian OS event-handling is usually managed in one thread, which runs a single active scheduler
- the active scheduler holds a set of active objects for that thread, each of which encapsulates an associated asynchronous service provider
- each thread has an associated request semaphore; when an asynchronous function completes, it generates an event by calling `RequestComplete()` on the requesting thread, which signals its semaphore and is detected by the active scheduler
- the active scheduler calls the `RunL()` event handler method of the active object associated with the completion event.

In the previous chapter, I used example code for an active object wrapper over an `RTimer` to illustrate the main points. You may find it useful to refer to that example throughout this chapter.

9.1 Active Object Basics

All active objects must derive from class `CActive`. On construction, each active object is assigned a priority value, which is set through a call to the base class constructor. In its constructor, the active object must also be added to the active scheduler through a call to `CActiveScheduler::Add()`. The active scheduler maintains a doubly-linked list of the active objects added to it, ordered by priority. When an active object is added to the active scheduler, it is added to that list in the appropriate position, according to its priority value.

As I described in Chapter 8, the active object encapsulates asynchronous functions (those that return immediately and complete at some later stage rather than returning only when the request has completed). On Symbian OS, asynchronous functions can be identified as those taking a `TRequestStatus` reference parameter into which the request completion status is posted. An object implementing such methods is usually known as an asynchronous service provider.

A typical active object class provides public "request issuer" methods for its clients to submit asynchronous requests. These pass on the requests to the encapsulated asynchronous service provider, passing by reference the `iStatus` member variable of the `CActive` class as the `TRequestStatus` parameter. Having issued the request, the issuer method must call `CActive::SetActive()` to set the `iActive` member to indicate that there is an outstanding request.

The service provider must set the value of the incoming `TRequestStatus` to `KRequestPending` (=`0x80000001`) before acting on the request. Upon completion, if the service provider is in the same thread as the requester, it calls `User::RequestComplete()`, passing the `TRequestStatus` and a completion result, typically one of the standard errors such as `KErrNone` or `KErrNotFound`, to indicate the success or otherwise of the request. `User::RequestComplete()` sets the value of `TRequestStatus` and generates a completion event in the requesting thread by signaling the thread's request semaphore. If the asynchronous service provider and the requester are in separate threads, the service provider must use an `RThread` object, representing a handle to the requesting thread, to complete the request. It should call `RThread::RequestComplete()` to post the completion code and notify the request semaphore.

While the request is outstanding, the requesting thread runs in the active scheduler's event processing loop. When it is not handling completion events, the active scheduler suspends the thread by calling `User::WaitForAnyRequest()`, which waits on a signal to the thread's request semaphore. When the asynchronous service provider completes a request, it signals the semaphore of the requesting thread as described above, and the active scheduler determines which active object should handle the completed request. It uses its priority-ordered list of active objects, inspecting each one in turn to determine whether it has a request outstanding. It does so by checking the `iActive` flag; if the object does indeed have an outstanding request, it then inspects its `TRequestStatus` member variable to see if it is set to a value other than `KRequestPending`. If so, this indicates that the active object is associated with a request that has completed and that its event handler code should be called.

Having found a suitable active object, the active scheduler clears the active object's `iActive` flag and calls its `RunL()` event handler. This method handles the event and may, for example, resubmit a request or generate an event on another object in the system. While this method is running, other events may be generated but `RunL()` is not pre-empted – it runs to completion before the active scheduler resumes control and determines whether any other requests have completed.

Once the `RunL()` call has finished, the active scheduler re-enters the event processing wait loop by issuing another `User::WaitForAnyRequest()` call. This checks the request semaphore and either suspends the thread (if no other requests have completed in the meantime) or returns immediately (if the semaphore indicates that other events were generated while the previous event handler was running) so the scheduler can repeat active object lookup and event handling.

Here's some pseudo-code which represents the basic actions of the active scheduler's event processing loop.

```
EventProcessingLoop()
    {
    // Suspend the thread until an event occurs
    User::WaitForAnyRequest();
    // Thread wakes when the request semaphore is signaled
    // Inspect each active object added to the scheduler,
    // in order of decreasing priority
    // Call the event handler of the first which is active & completed
    FOREVER
        {
        // Get the next active object in the priority queue
        if (activeObject->IsActive())
                && (activeObject->iStatus!=KRequestPending)
            {// Found an active object ready to handle an event
             // Reset the iActive status to indicate it is not active
            activeObject->iActive = EFalse;
            // Call the active object's event handler in a TRAP
            TRAPD(r, activeObject->RunL());
            if (KErrNone!=r)
                {// event handler left, call RunError() on active object
                r = activeObject->RunError();
                if (KErrNone!=r) // RunError() didn't handle the error,
                    Error(r);    // call CActiveScheduler::Error()
                }
            break; // Event handled. Break out of lookup loop & resume
```

```
        }
    } //   End of FOREVER loop
}
```

If a single request has completed on the thread in the interim, the active scheduler performs lookup and calls the appropriate event handler on that active object. If more than one request has completed in that time, the active scheduler calls the event handler for the highest priority active object. It follows that, if multiple events are generated in close succession while another event is being handled, those events may not be handled in the sequence in which they occurred because the active object search list is ordered by priority to support responsive event-handling.

Normally, active object code should be designed so the priority does not matter, otherwise the system can become rather delicate and be thrown off balance by minor changes or additional active objects on the thread. However, to be responsive, say for user input, it is sometimes necessary to use a higher priority value. Long-running, incremental tasks, on the other hand, should have a lower priority than standard since they are designed to use idle processor time (as I'll describe later in this chapter).

It's important to understand that the priority value is only an indication of the order in which the active scheduler performs lookup and event-handling when multiple events have completed. In contrast to the priority values of threads used by the kernel scheduler, it does not represent an ability to pre-empt other active objects. Thus, if you assign a particular active object a very high priority, and it completes while a lower-priority active object is handling an event, no pre-emption occurs. The `RunL()` of the lower-priority object runs to completion, regardless of the fact that it is "holding up" the handler for the higher-priority object.

On Symbian OS, you cannot use active object priorities to achieve a guaranteed response time; for this you must use the pre-emptive scheduling associated with threads[1], which is described in Chapter 10.

If you have a large number of active objects in a single thread which complete often, they "compete" for their event handler to be run by the active scheduler. If some of the active objects have high priorities and receive frequent completion events, those with lower priorities wait indefinitely until the active scheduler can call their `RunL()` methods. In effect, it's possible to "hang" lower-priority active objects by adding them to an active scheduler upon which a number of high-priority active objects are completing.

[1] The new hard real-time kernel in Symbian OS 8.0 can commit to a particular response time. On earlier versions of Symbian OS, the kernel has soft real-time capabilities and cannot make such guarantees.

An active object's priority is only an indication of the order in which the active scheduler performs lookup and event-handling. It does not reflect an ability to pre-empt other active objects.

9.2 Responsibilities of an Active Object

Figure 9.1 illustrates the roles and actions of the active scheduler, active object and the asynchronous service provider. It extends Figure 8.1.

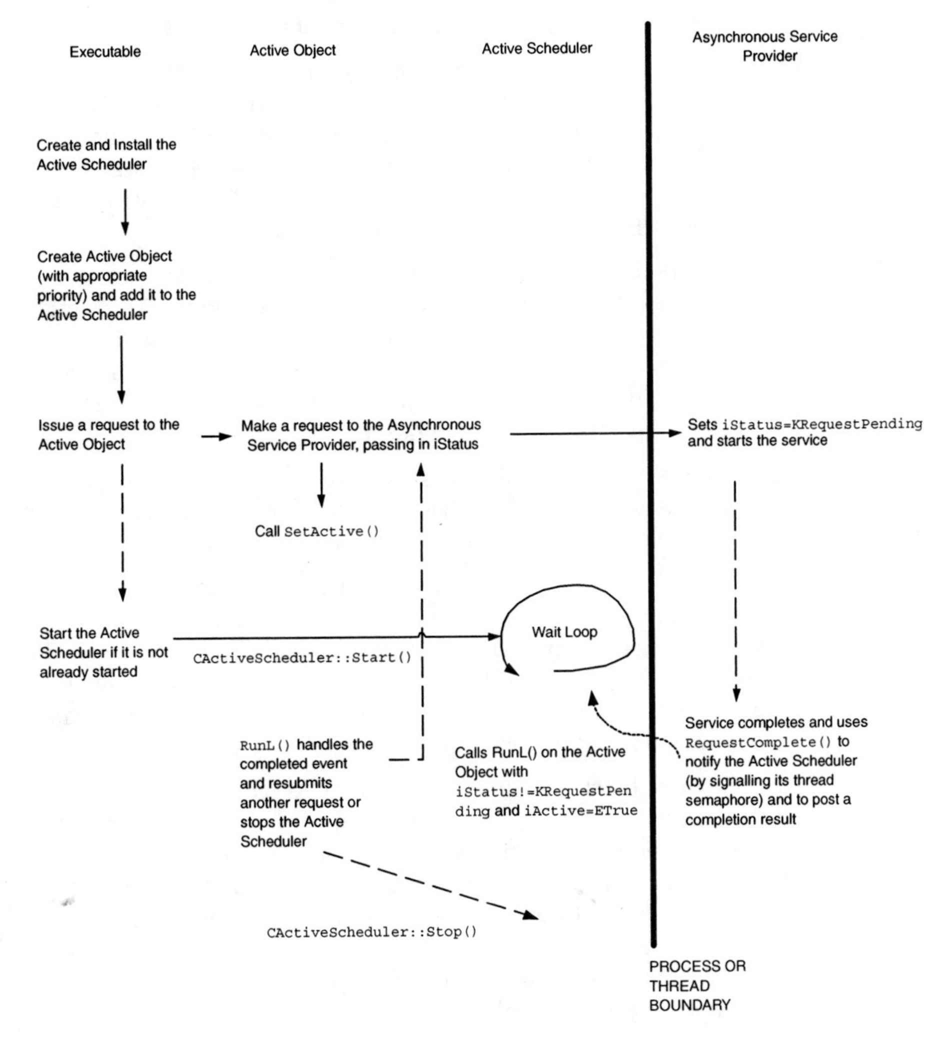

Figure 9.1 Roles and actions of the active scheduler, an active object and an asynchronous service provider

It looks complex, but I'll explain how it all fits together throughout this chapter and you'll probably want to refer back to it later.

The following list summarizes the responsibilities of an active object:

- As I described in Chapter 8, the priority of an active object must be set on construction. The priority generally defaults to `EPriorityStandard` (=0, from class `CActive`) or `EActivePriorityDefault` (=0 if using the `TActivePriority` enumeration defined in `coemain.h` for use with application code). This is the standard priority for an active object and should be used unless there is a good reason to set its priority to some other value, for example to `EActivePriorityWsEvents` (=100) for handling user input responsively.
- An active object provides at least one method for clients to initiate requests to its encapsulated asynchronous service provider. The active object always passes its own `iStatus` object to the asynchronous function, so does not need to include a `TRequestStatus` reference among the parameters to the request issuer method unless it is acting as a secondary provider of asynchronous services.
- After submitting a request to an asynchronous service provider, the active object must call `SetActive()` upon itself. This sets the `iActive` flag, which indicates an outstanding request. This flag is used by the active scheduler upon receipt of an event and by the base class upon destruction, to determine whether the active object can be removed from the active scheduler.
- An active object should only submit one asynchronous request at a time. The active scheduler has no way of managing event-handling for multiple requests associated with one active object.
- An active object should pass its `iStatus` object to an asynchronous service function. It should not reuse that object until the asynchronous function has completed and been handled. The active scheduler inspects the `TRequestStatus` of each active object to determine its completion state and the event-handling code uses the value it contains to ascertain the completion result of the function.
- An active object must implement the pure virtual methods `RunL()` and `DoCancel()` declared in the `CActive` base class. Neither method should perform lengthy or complex processing, to avoid holding up event handling in the entire thread. This is particularly important in GUI applications where all user interface code runs in the same thread. If any single `RunL()` is particularly lengthy, the user interface will be unresponsive to input and will "freeze" until that event handler has completed.

- An active object must ensure that it is not awaiting completion of a pending request when it is about to be destroyed. As destruction removes it from the active scheduler, later completion will generate an event for which there is no associated active object. To prevent this, `Cancel()` should be called in the destructor of an active object. The destructor of the `CActive` base class checks that there is no outstanding request before removing the object from the active scheduler and raises an `E32USER-CBASE 40` panic if there is, to highlight the programming error. The base class destructor cannot call `Cancel()` itself because that method calls the derived class implementation of `DoCancel()` – and, of course, C++ dictates that the derived class has already been destroyed by the time the base class destructor is called.
- Objects passed to the asynchronous service provider by the issuer methods must have a lifetime at least equal to the time taken to complete the request. This makes sense when you consider that the provider may use those objects until it is ready to complete the request, say if it is retrieving and writing data to a supplied buffer. This requirement means that parameters supplied to the provider should usually be allocated on the heap (very rarely, they may be on a stack frame that exists for the lifetime of the entire program). In general, parameters passed to the asynchronous service provider should belong to the active object, which is guaranteed to exist while the request is outstanding.
- If a leave can occur in `RunL()`, the class should override the default implementation of the virtual `RunError()` method to handle it. `RunError()` was added to `CActive` in Symbian OS v6.0 to handle any leaves that occur in the `RunL()` event handler. If a leave occurs, the active scheduler calls the `RunError()` method of the active object, passing in the leave code. `RunError()` should return `KErrNone` to indicate that it has handled the leave, say by cleaning up or resetting the state of the active object. The default implementation, `CActive::RunError()`, does not handle leaves and indicates this by simply returning the leave code passed in.

9.3 Responsibilities of an Asynchronous Service Provider

An asynchronous service provider has the following responsibilities:

- Before beginning to process each request, the provider must set the incoming `TRequestStatus` value to `KRequestPending` to indicate to the active scheduler that a request is ongoing.

- When the request is complete, the provider must set the `TRequestStatus` value to a result code other than `KRequestPending` by calling the appropriate `RequestComplete()` method from the `RThread` or `User` class.

- The asynchronous service provider must only call `RequestComplete()` once for each request. This method generates an event in the requesting thread to notify it of completion. Multiple completion events on a single active object result in a stray signal panic. Completion may occur normally, because of an error condition or because the client has cancelled an outstanding request. If the client calls `Cancel()` on a request *after* it has completed, the asynchronous service provider must not complete it again and should simply ignore the cancellation request. This is discussed further in Sections 9.8 and 9.9.

- The provider must supply a corresponding cancellation method for each asynchronous request; this should complete an outstanding request *immediately*, posting `KErrCancel` into the `TRequestStatus` object associated with the initial request.

9.4 Responsibilities of the Active Scheduler

The active scheduler has the following responsibilities:

- Suspending the thread by a call to `User::WaitForAnyRequest()`. When an event is generated, it resumes the thread and inspects the list of active objects to determine which has issued the request that has completed and should be handled.

- Ensuring that each request is handled only once. The active scheduler should reset the `iActive` flag of an active object before calling its handler method. This allows the active object to issue a new request from its `RunL()` event handler, which results in `SetActive()` being called (which would panic if the active object was still marked active from the previous request).

- Placing a `TRAP` harness around `RunL()` calls to catch any leaves occurring in the event-handling code. If the `RunL()` call leaves, the active scheduler calls `RunError()` on the active object initially. If the leave is not handled there, it passes the leave code to `CActiveScheduler::Error()`, described in more detail shortly.

- Raising a panic (`E32USER-CBASE 46`) if it receives a "stray signal". This occurs when the request semaphore has been notified of an event, but the active scheduler cannot find a "suitable" active object (with iActive set to ETrue and a TRequestStatus indicating that it has completed).

9.5 Starting the Active Scheduler

Once an active scheduler has been created and installed, its event processing wait loop is started by a call to the static CActive-Scheduler::Start() method. Application programmers do not have to worry about this, since the CONE framework takes care of managing the active scheduler. If you are writing server code, or a simple console application, you have to create and start the active scheduler for your server thread, which can be as simple as follows:

```
CActiveScheduler* scheduler = new(ELeave) CActiveScheduler;
CleanupStack::PushL(scheduler);
CActiveScheduler::Install(scheduler);
```

The call to Start() enters the event processing loop and does not return until a corresponding call is made to CActive-Scheduler::Stop(). Thus, before the active scheduler is started, there must be at least one asynchronous request issued, via an active object, so that the thread's request semaphore is signaled and the call to User::WaitForAnyRequest() completes. If no request is outstanding, the thread simply enters the wait loop and sleeps indefinitely.

As you would expect, the active scheduler is stopped by a call to CActiveScheduler::Stop(). When that enclosing function returns, the outstanding call to CActiveScheduler::Start() also returns. Stopping the active scheduler breaks off event handling in the thread, so it should only be called by the main active object controlling the thread.

9.6 Nesting the Active Scheduler

I've already noted that an event-handling thread has a single active scheduler. However, it is possible, if unusual, to nest other calls to CActiveScheduler::Start(), say within a RunL() event-handling method. The use of nested active scheduler loops is generally discouraged but can be useful if a call should appear to be synchronous, while actually being asynchronous ("pseudo-synchronous"). A good example is a RunL() event handler that requires completion of an asynchronous request to another active object in that thread. The RunL() call cannot be pre-empted, so it must instead create a nested wait loop by calling CActiveScheduler::Start(). This technique is used in modal Uikon "waiting" dialogs.

Each call to CActiveScheduler::Start() should be strictly matched by a corresponding call to CActiveScheduler::Stop() in an appropriate event handler. Before employing such a technique you must be careful to test your code thoroughly to ensure that the nesting is controlled under both normal and exceptional conditions. The use of nested active scheduler event-processing loops can introduce subtle bugs, particularly if more than one nested loop is used concurrently in the same thread. For example, if a pair of independent components both nest active scheduler loops, their calls to Start() and Stop() must be carefully interleaved if one component is to avoid stopping the loop of the other's nested active scheduler.

The complexity that results from nesting active scheduler processing loops means that Symbian does not recommend this technique. However, where the use of nested active scheduler loops is absolutely unavoidable, releases of Symbian OS from v7.0s onwards have introduced the CActiveSchedulerWait class to provide nesting "levels" that match active scheduler Stop() calls to the corresponding call to Start().

9.7 Extending the Active Scheduler

CActiveScheduler is a concrete class and can be used "as is", but it can also be subclassed. It defines two virtual functions which may be extended: Error() and WaitForAnyRequest().

By default, the WaitForAnyRequest() function simply calls User::WaitForAnyRequest(), but it may be extended, for example to perform some processing before or after the wait. If the function is re-implemented, it must either call the base class function or make a call to User::WaitForAnyRequest() directly.

I described earlier how if a leave occurs in a RunL() event handler, the active scheduler passes the leave code to the RunError() method of the active object. If this method cannot handle the leave, it returns the leave code and the active scheduler passes it to its own Error() method. By default, this raises a panic (E32USER-CBASE 47), but it may be extended in a subclass to handle the error, for example by calling an error resolver to obtain the textual description of the error and displaying it to the user or logging it to file.

If your active object code is dependent upon particular specializations of the active scheduler, bear in mind that it will not be portable to run in other threads managed by more basic active schedulers. Furthermore, any additional code added to extend the active scheduler should be straightforward and you should avoid holding up event-handling in the entire thread by performing complex or slow processing.

9.8 Cancellation

Every request issued by an active object must complete exactly once. It can complete normally or complete early as a result of an error or a call to `Cancel()`. Let's first examine what happens in a call to `CActive::Cancel()` and return to the other completion scenarios later.

`CActive::Cancel()` first determines if there is an outstanding request and, if so, it calls the `DoCancel()` method, a pure virtual function in `CActive`, implemented by the derived class (which should *not* override the non-virtual base class `Cancel()` method). `DoCancel()` does not need to check if there is an outstanding request; if there is no outstanding request, `Cancel()` does not call it. The encapsulated asynchronous service provider should provide a method to cancel an outstanding request and `DoCancel()` should call this method.

`DoCancel()` can include other processing, but it should not leave or allocate resources and it should not carry out any lengthy operations. This is because `Cancel()` is itself a synchronous function which does not return until both `DoCancel()` has returned and the original asynchronous request has completed. That is, having called `DoCancel()`, `CActive::Cancel()` then calls `User::WaitForRequest()`, passing in a reference to its `iStatus` member variable. It is blocked until the asynchronous service provider posts a result (`KErrCancel`) into it, which should happen immediately, as described above.

The cancellation event is thus handled by the `Cancel()` method of the active object rather than by the active scheduler.

Finally, `Cancel()` resets the `iActive` member of the active object to reflect that there is no longer an asynchronous request outstanding.

The `Cancel()` method of the `CActive` base class performs all this generic cancellation code. When implementing a derived active object class, you only need to implement `DoCancel()` to call the appropriate cancellation function on the asynchronous service provider and perform any cleanup necessary. You most certainly should *not* call `User::WaitForRequest()`, since this will upset the thread semaphore count. Internally, the active object must not call the protected `DoCancel()` method to cancel a request; it should call `CActive::Cancel()`, which invokes `DoCancel()` and handles the resulting cancellation event.

When an active object request is cancelled by a call to `Cancel()`, the `RunL()` event handler does not run. This means that any post-cancellation cleanup must be performed in `DoCancel()` rather than in `RunL()`.

9.9 Request Completion

At this point, we can summarize the ways in which a request issued from an active object to an asynchronous service provider can complete:

- The request is issued to the asynchronous service provider by the active object. Some time later, the asynchronous service provider calls `User::RequestComplete()` which generates a completion event and passes back a completion result. The active scheduler detects the completion event, resumes the thread and initiates event handling on the highest priority active object that has `iActive` set to `ETrue` and `iStatus` set to a value other than `KRequestPending`. This is a normal case, as described in the walkthrough above, although the completion result may not reflect a successful outcome.
- The asynchronous request cannot begin, for example if invalid parameters are passed in or insufficient resources are available. The asynchronous service provider should define a function that neither leaves nor returns an error code (it should typically return `void`). Thus, under these circumstances, the request should complete immediately, posting an appropriate error into the `TRequestStatus` object passed into the request function.
- The request is issued to the asynchronous service provider and `Cancel()` is called on the active object before the request has completed. The active object calls the appropriate cancellation function on the asynchronous service provider, which should terminate the request immediately. The asynchronous service provider should complete the request with `KErrCancel` as quickly as possible, because `CActive::Cancel()` blocks until completion occurs.
- The request is issued to the asynchronous service provider and `Cancel()` is called on the active object some time after the request has completed. This occurs when the completion event has occurred but is yet to be processed by the active scheduler. The request appears to be outstanding to the active object framework, if not to the asynchronous service provider, which simply ignores the cancellation call. `CActive::Cancel()` discards the normal completion result.

9.10 State Machines

An active object class can be used to implement a state machine to perform a series of actions in an appropriate sequence, without requiring client code to make multiple function calls or understand the logic of the sequence. The example below is of an active object class, CStateMachine, which has a single request method SendTranslatedData(). This retrieves the data, converts it in some way and sends it to another location. The method takes the location of a data source, the destination and a TRequestStatus which is stored and used to indicate to the caller when the series of steps has completed. CStateMachine encapsulates an object of CServiceProvider class which provides the methods necessary to implement SendTranslatedData(). This class acts as an asynchronous service provider for the active object. Each asynchronous method takes a reference to a TRequestStatus object and has a corresponding Cancel() method.

The state machine class has an enumeration which represents the various stages required for SendTranslatedData() to succeed. It starts as CStateMachine::EIdle and must be in this state when the method is called (any other state indicates that a previous call to the method is currently outstanding). Having submitted a request by making the first logical call to CServiceProvider::GetData(), the method changes the iState member variable to reflect the new state (CStateMachine::EGet) and calls SetActive(). When it has finished, GetData() generates a completion event and the active scheduler, at some later point, invokes the CStateMachine::RunL() event handler. This is where the main logic of the state machine is implemented. You'll see from the example that it first checks whether an error has occurred and, if so, it aborts the rest of the sequence and notifies the client. Otherwise, if the previous step was successful, the handler calls the next method in the sequence and changes its state accordingly, again calling SetActive(). This continues, driven by event completion and the RunL() event handler.

For clarity, in the example code below, I've only shown the implementation of functions which are directly relevant to the state machine:

```
// Provides the "step" functions
class CServiceProvider : public CBase
    {
public:
    static CServiceProvider* NewL();
    ~CServiceProvider() {};
public:
    void GetData(const TDesC& aSource, HBufC8*& aData,
            TRequestStatus& aStatus);
    void CancelGetData();
    TInt TranslateData(TDes8& aData);
    void SendData(const TDesC& aTarget, const TDesC8& aData,
            TRequestStatus& aStatus);
    void CancelSendData();
protected:
```

```
    CServiceProvider(){};
    };

void CServiceProvider::GetData(const TDesC& aSource, HBufC8*& aData,
        TRequestStatus& aStatus)
    {
    aStatus = KRequestPending;

    // Retrieves data from aSource using the asynchronous overload of
    // RFile::Read() and writing to aData (re-allocating it if
    // necessary). aStatus is completed by the file server when the
    // read has finished
    ...
    }

void CServiceProvider::CancelGetData() {...}

TInt CServiceProvider::TranslateData(TDes8& aData)
    {// Synchronously translates aData & writes into same descriptor
    ...
    return (translationResult);
    }

void CServiceProvider::SendData(const TDesC& aTarget, const TDesC8&
aData, TRequestStatus& aStatus)
    {
    aStatus = KRequestPending;

    // Writes data to aTarget using the asynchronous overload of
    // RFile::Write(), which completes aStatus
    ...
    }

void CServiceProvider::CancelSendData() {...}

class CStateMachine : public CActive
    {
public:
    ~CStateMachine();
    static CStateMachine* NewLC();
    void SendTranslatedData(const TDesC& aSource, const TDesC& aTarget,
        TRequestStatus&);
protected:
    enum TState { EIdle, EGet, ETranslate, ESend};
protected:
    CStateMachine();
    void InitializeL(const TDesC& aTarget);
    void Cleanup();
protected:
    virtual void DoCancel(); // Inherited from CActive
    virtual void RunL();
    // The following base class method is not overridden because
    // RunL() cannot leave
    // virtual TInt RunError(TInt aError);
private:
    CServiceProvider* iService;
    TState iState;
private:
```

```
    HBufC* iTarget;
    HBufC8* iStorage;
    TRequestStatus* iClientStatus;
    };

CStateMachine::CStateMachine()
: CActive(EPriorityStandard) {CActiveScheduler::Add(this);}

CStateMachine::~CStateMachine()
    {
    Cancel();
    Cleanup();
    }

void CStateMachine::InitializeL(const TDesC& aTarget)
    {
    // Store this to pass to CServiceProvider later
    iTarget = aTarget.AllocL();
    // To store retrieved data
    iStorage = HBufC8::NewL(KStandardDataLen);
    }

void CStateMachine::Cleanup()
    {// Pointers are NULL-ed because this method is called outside
     // the destructor
    iState = EIdle;
    delete iTarget;
    iTarget = NULL;
    delete iStorage;
    iStorage = NULL;
    }

const TInt KStandardDataLen = 1024;

    // Starts the state machine
void CStateMachine::SendTranslatedData(const TDesC& aSource,
        const TDesC& aTarget, TRequestStatus& aStatus)
    {
    __ASSERT_ALWAYS(!IsActive(), User::Panic(KExPanic, KErrInUse));
    ASSERT(EIdle==iState);
    // Store the client request status to complete later
    iClientStatus = &aStatus;
    iClientStatus = KRequestPending;

    TRAPD(r, InitializeL(aTarget);
    if (KErrNone!=r)
        {// Allocation of iTarget of iStorage failed
        Cleanup(); // Destroys successfully allocated member data
        User::RequestComplete(iClientStatus, r);
        }
    else
        {
        iService->GetData(aSource, iStorage, iStatus);
        iState = EGet;
        SetActive();
        }
    }
```

```
// The state machine is driven by this method
void CStateMachine::RunL()
    {// Inspects result of completion and iState
     // and submits next request (if required)
    ASSERT(EIdle!=iState);

    if (KErrNone!=iStatus.Int())
        {// An error - notify the client and stop the state machine
        User::RequestComplete(iClientStatus, iStatus.Int());
        Cleanup();
        }
    else
        {
        if (EGet==iState)
            {// Data was retrieved, now translate it synchronously
            TPtr8 ptr(iStorage->Des());
            iService->TranslateData(ptr);
            iState = ETranslate;
            // Self completion - described later
            TRequestStatus* stat = &iStatus;
            User::RequestComplete(stat, r);
            SetActive();
            }
        else if (ETranslate==iState)
            {// Data was translated, now send it asynchronously
            TInt r = iService->SendData(*iTarget, *iStorage, iStatus);
            iState = ESend;
            SetActive();
            }
        else
            {// All done, notify the caller
            ASSERT(ESend==iState);
            User::RequestComplete(iClientStatus, iStatus.Int());
            Cleanup();
            }
        }
    }

void CStateMachine::DoCancel()
    {
    if (iService)
        {
        if (CStateMachine::EGet = =iState)
            {
            iService->CancelGetData();
            }
        else if (CStateMachine::ESend = =iState)
            {
            iService->CancelSendData();
            }
        }
    if (iClientStatus)
        {// Complete the caller with KErrCancel
        User::RequestComplete(iClientStatus, KErrCancel);
        }
    Cleanup();
    }
```

In effect, CStateMachine maintains a series of outstanding requests to the service provider in RunL(), rather than making a single call. This example is quite straightforward because there are only three active states, but the code logic here can potentially be far more complex. The DoCancel() method of CStateMachine must also have some state-related logic so that the correct method on CServiceProvider is cancelled. The states and transitions are illustrated in Figure 9.2.

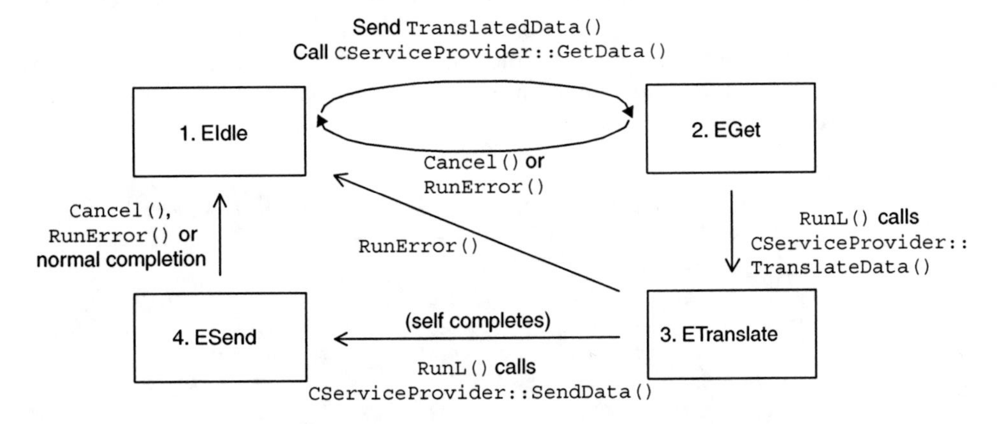

Figure 9.2 Internal states of CStateMachine

In this example the service provider functions called by the state machine are a mixture of synchronous (TranslateData()) and asynchronous (GetData() and SendData()) functions. The synchronous method uses self-completion to simulate an asynchronous completion event, which is discussed further in the following section.

9.11 Long-Running Tasks

Besides encapsulating asynchronous service providers, active objects can also be used to implement long-running tasks which would otherwise need to run in a lower-priority background thread.

To be suitable, the task must be divisible into multiple short increments, for example, preparing data for printing, performing background recalculations and compacting the database. The increments are performed in the event handler of the active object. For this reason, they must be short enough for event handling in the thread to continue to be responsive, because RunL() cannot be pre-empted once it is running.

The active object should be assigned a low priority such as CActive::TPriority::EPriorityIdle (=-100), which determines that a task increment only runs when there are no other events to handle, i.e. in idle time. If the task consists of a number of different steps, the active object must track the progress as series of states, implementing it using a state machine as described above.

The active object drives the task by generating its own events to invoke its event handler. Instead of calling an asynchronous service provider, it completes itself by calling `User::RequestComplete()` on its own `iStatus` object and calls `SetActive()` on itself so the active scheduler calls its event handler. In this way it continues to resubmit requests until the entire task is complete. A typical example is shown in the sample code, where I've shown all the relevant methods in the class declarations but only the implementations relevant to this discussion. I've also omitted error handling for clarity:

```
// This class has no dependence on the active object framework
class CLongRunningCalculation : public CBase
    {
public:
    static CLongRunningCalculation* NewL();
    TBool StartTask();  // Initialization before starting the task
    TBool DoTaskStep(); // Performs a short task step
    void EndTask();     // Destroys intermediate data
    ...
    };

    TBool CLongRunningCalculation::DoTaskStep()
    {// Do a short task step, returning
     // ETrue if there is more of the task to do
     // EFalse if the task is complete
    ... // Omitted
    }

_LIT(KExPanic, "CActiveExample");

class CBackgroundRecalc : public CActive
    {
public:
    ... // NewL(), destructor etc are omitted for clarity
public:
    void PerformRecalculation(TRequestStatus& aStatus);
protected:
    CBackgroundRecalc();
    void ConstructL();
    void Complete();
    virtual void RunL();
    virtual void DoCancel();
private:
    CLongRunningCalculation* iCalc;
    TBool iMoreToDo;
    TRequestStatus* iCallerStatus; // To notify caller on completion
    };

CBackgroundRecalc::CBackgroundRecalc()
    : CActive(EPriorityIdle)  //  Low priority task
    { CActiveScheduler::Add(this); }
```

```
    // Issues a request to initiate a lengthy task
void CBackgroundRecalc::PerformRecalculation(TRequestStatus& aStatus)
    {
    iCallerStatus = &aStatus;
    *iCallerStatus = KRequestPending;
    __ASSERT_ALWAYS(!IsActive(), User::Panic(KExPanic, KErrInUse));
    iMoreToDo = iCalc->StartTask(); //  iCalc initializes the task
    Complete();                    //  Self-completion to generate an event
    }

void CBackgroundRecalc::Complete()
    {// Generates an event on itself by completing on iStatus
    TRequestStatus* status = &iStatus;
    User::RequestComplete(status, KErrNone);
    SetActive();
    }

    // Performs the background task in increments
void CBackgroundRecalc::RunL()
    {// Resubmit request for next increment of the task or stop
    if (!iMoreToDo)
        {// Allow iCalc to cleanup any intermediate data
        iCalc->EndTask();
        // Notify the caller
        User::RequestComplete(iCallerStatus, iStatus.Int());
        }
    else
        {// Submit another request and self-complete to generate event
        iMoreToDo = iCalc->DoTaskStep();
        Complete();
        }
    }

void CBackgroundRecalc::DoCancel()
    {// Give iCalc a chance to perform cleanup
    if (iCalc)
        iCalc->EndTask();

    if (iCallerStatus) // Notify the caller
        User::RequestComplete(iCallerStatus, KErrCancel);
    }
```

If you are designing an API for a long-running task, to make it useful with this pattern, it is a good idea to provide methods which split the task into increments, as in the example above. `StartTask()`, `DoTaskStep()` and `EndTask()` perform small, discrete chunks of the task and can be called directly by the `RunL()` method of the low-priority active object. The API can then also be reused by code which implements long-running tasks differently, since it will have no dependence on the active object itself. Besides portability, distancing the active object model makes it more straightforward to focus on the implementation of the long-running task itself.

Of course, one disadvantage of this approach is that some tasks cannot be broken down into short steps. Another is that if you implement a number of low-priority active objects for long-running tasks in the same thread, you will need to work out how best to run them together and write the necessary low-priority scheduling code yourself.

The use of a background thread for long-running tasks is fairly straightforward. The code for the task can be written without worrying about yielding the CPU, since the kernel schedules it when no higher-priority thread needs to run, pre-empting it again when a more important thread needs access to system resources. However, as with all multi-threaded code, any shared resources must be protected against illegal concurrent access by synchronization objects and, on Symbian OS, the overhead of running multiple threads is significantly higher than that for multiple active objects running in a single thread. You should prefer low-priority active objects for long-running tasks, except for cases where the task cannot be split into convenient increments. The next chapter illustrates how to use a separate thread to perform a long-running task which is wrapped with code for an active object.

9.12 Class `CIdle`

`CIdle` derives from `CActive` and is a useful class which wraps the active object basics such as implementing `RunL()` and `DoCancel()`. The wrapper allows you to focus solely on implementing the code to run the incremental task without worrying about the active object code.

```
class CIdle : public CActive
    {
public:
    IMPORT_C static CIdle* New(TInt aPriority);
    IMPORT_C static CIdle* NewL(TInt aPriority);
    IMPORT_C ~CIdle();
    IMPORT_C void Start(TCallBack aCallBack);
protected:
    IMPORT_C CIdle(TInt aPriority);
    IMPORT_C void RunL();
    IMPORT_C void DoCancel();
protected:
    TCallBack iCallBack;
};
```

The `CIdle` object should be created with a low or idle priority and the long-running task initiated by a call to `Start()`, passing a callback function to perform an increment of the task. The `TCallback` object encapsulates a pointer to a callback function which takes an argument of type `TAny*` and returns an integer. The callback function manages the task increments and can be a local or a static member function. It should keep track of the task progress, returning `ETrue` if further steps are necessary and EFalse when it is complete. In much the same way as the incremental task shown above, the RunL() event handler which calls the TCallback object is only called during idle time. Furthermore, it will not be pre-empted while it is running. As long as the callback function indicates that further steps of the task are required, CIdle::RunL() resubmits requests by completing on its own iStatus object and calling SetActive().

Here's some example code for another background recalculation task class. In fact, I've slightly reworked the class from the example above to be driven by CIdle. You'll notice that there's no "boilerplate" active object code required, unlike the code in the CBackgroundRecalc class, because CIdle provides that functionality.

```
class CLongRunningCalculation : public CBase
    {
public:
    static CLongRunningCalculation* NewLC(TRequestStatus& aStatus);
protected:
    static TInt TaskStep(TAny* aLongCalc); // Callback function
protected:
    void StartTask();    // Initialization before starting the task
    TBool DoTaskStep(); // Does a step of the task
    void EndTask();      // Destroys intermediate data
protected:
    CLongRunningCalculation(TRequestStatus& aStatus);
private:
    ...
    TBool iMoreToDo; // Flag tracks the progress of the task
    // To notify the caller when the calc is complete
    TRequestStatus* iCallerStatus;
    };

CLongRunningCalculation* CLongRunningCalculation::NewLC(TRequestStatus&
aStatus)
    {
    CLongRunningCalculation* me = new (ELeave)
    CLongRunningCalculation(aStatus);
    CleanupStack::PushL(me); // ... 2nd phase construction code omitted
    return (me);
    }

CLongRunningCalculation::CLongRunningCalculation(TRequestStatus&
aStatus)
: iCallerStatus(&aStatus) {}

TBool CLongRunningCalculation::DoTaskStep()
    {// Does a short task step, returning ETrue if it has more to do
    ... // Omitted
    }

void CLongRunningCalculation::StartTask()
    {// Prepares the task
    iMoreToDo = ETrue;
    }
```

```
// Error handling omitted
void CLongRunningCalculation::EndTask()
    {// Performs cleanup when the task has completed
    ASSERT(!iMoreToDo);
    ...
    User::RequestComplete(iCallerStatus, KErrNone);
    }

TInt CLongRunningCalculation::TaskStep(TAny* aLongCalc)
    {
    ASSERT(aLongCalc);
    CLongRunningCalculation* longCalc =
            static_cast<CLongRunningCalculation*>(aLongCalc);
    if (!longCalc->iMoreToDo)
        longCalc->EndTask();
    else
        longCalc->iMoreToDo = longCalc->DoTaskStep();

    return (longCalc->iMoreToDo);
    }
```

Code which uses the idle object will look something like the following example, which creates the `CIdle` object and a `CLongRunningCalculation` object that drives the task:

```
CIdle* idle = CIdle::NewL(EPriorityIdle);
CleanupStack::PushL(idle);

// Create the object that runs the task, passing in a TRequestStatus&
CLongRunningCalculation* longCalc =
        CLongRunningCalculation::NewLC(iStatus);
TCallBack callback(CLongRunningCalculation::TaskStep, longCalc);
idle->Start(callback);
... // Completion of the task completes iStatus
```

9.13 Class `CPeriodic`

`CPeriodic` is another useful `CActive`-derived class for running incremental task steps. It uses a timer object to generate regular timer events, handling a single step of the task for each timer event. `CPeriodic` is useful for performing regular task steps such as flashing a cursor or for controlling time-outs.

Like `CIdle`, `CPeriodic` is initialized with a priority value and the task is initiated by a call to `Start()`, passing settings for the timer as well as a callback to perform increments of the task. When the timer period elapses, the `CPeriodic` object generates an event that is detected by the active scheduler. When there are no active objects of higher priority requiring event handling, the active scheduler calls the `RunL()` event handler of the CPeriodic object, which in turn calls its task-processing callback function. Thus, the callback may not be as exactly regular as the periodic value passed to the Start() request. If the timer completes but other, higher-priority, active objects have completed events, they are processed first. Alternatively, the RunL() method of another active object may be running when the timer elapses and a running active object cannot be pre-empted even if it has a lower priority than the CPeriodic object.

The response of signal and callback processing can be improved by ensuring that all active objects in the thread perform short, efficient RunL() methods and that the CPeriodic object has a higher priority than other active objects.

9.14 Common Mistakes

I've described the dos and don'ts of active objects fairly comprehensively in this chapter and the previous one. The most commonly encountered problem when writing active object code is the infamous "stray signal" panic (E32USER-CBASE 46), which occurs when the active scheduler receives a completion event but cannot find an active object to handle it (i.e. one which is currently active and has a completed iStatus result, indicated by a value other than KRequestPending). Stray signals can arise for the following reasons:

- CActiveScheduler::Add() was not called when the active object was constructed
- SetActive() was not called following the submission of a request to the asynchronous service provider
- the asynchronous service provider completed the TRequestStatus of an active object more than once – either because of a programming error in the asynchronous service provider (for example, when an already-completed request is cancelled) or because more than one request was submitted simultaneously on the same active object.

If a stray signal occurs from one of your active objects, it is worth checking against each of these.

Over the course of Chapters 8 and 9, I have described the cooperative multitasking nature of active objects, which means that an active object's RunL() event handler cannot be pre-empted by any other in that thread, except by using nested active scheduler loops, which are strongly discouraged. In consequence, when using active objects for event handling in, for example, a UI thread, their event-handler methods must be kept short to keep the UI responsive.

No active object should have a monopoly on the active scheduler that prevents other active objects from handling events. Active objects should be "cooperative". This means you should guard against:

- writing lengthy `RunL()` or `DoCancel()` methods
- repeatedly resubmitting requests
- assigning your active objects a higher priority than is necessary.

Stray signals can arise if:

- **the active object is not added to the active scheduler**
- **`SetActive()` is not called following the submission of a request**
- **the active object is completed more than once for any given request.**

9.15 Summary

While the previous chapter gave a high-level overview of active objects on Symbian OS, this chapter focused on active objects in detail, walking through the roles, responsibilities and interactions between the active scheduler, active objects and the asynchronous service providers they encapsulate.

It contained the detail necessary to write good active object or asynchronous service provider code and to extend the active scheduler. Example code illustrated the use of active objects for state machines and for implementing background step-wise tasks, either using active objects directly or through the `CIdle` wrapper class.

At this level of detail, active objects can seem quite complex and this chapter should mostly be used for reference until you start to work directly with complex active object code.

10

Symbian OS Threads and Processes

Don't disturb my circles!

Said to be the last words of Archimedes who was drawing geometric figures in the dust and became so absorbed that he snapped at a Roman soldier. The soldier became enraged, pulled out his sword and killed him

Chapters 8 and 9 discussed the role of active objects in multitasking code on Symbian OS. Active objects are preferred to threads for this role because they were designed specifically to suit the resource-limited hardware upon which Symbian OS runs. Multithreaded code has significantly higher run-time requirements compared to active objects: for example, a context switch between threads may be an order of magnitude slower than a similar switch between active objects running in the same thread.[1] Threads tend to have a larger size overhead too, typically requiring 4 KB kernel-side and 8 KB user-side for the program stack, compared to active objects, which generally occupy only the size of the C++ object (often less than 1 KB).

One of the main differences between multitasking with threads and active objects is the way in which they are scheduled. Active objects multitask cooperatively within a thread and, once handling an event, an active object cannot be pre-empted by the active scheduler in which it runs. Threads on Symbian OS are scheduled pre-emptively by the kernel.

Pre-emptive scheduling of threads means that data shared by threads must be protected with access synchronization primitives such as mutexes or semaphores. However, despite this additional complexity, pre-emptive scheduling is sometimes necessary. Consider the case of two active objects, one assigned a high priority because it handles user-input events

[1] A context switch between threads running in the same process requires the processor registers of the running thread to be stored, and the state of the thread replacing it to be restored. If a reschedule occurs between threads running in two separate processes, the address space accessible to the thread, the process context, must additionally be stored and restored.

and another with a lower priority, which performs increments of a long-running task when no other active object's event handler is running on the thread. If the event handler of the lower-priority active object happens to be running when an event for the higher-priority active object completes, it will continue to run to completion. No pre-emption will occur and the low-priority active object effectively "holds up" the handler of the high-priority object. This can make the user interface sluggish and unresponsive.

Typically, on Symbian OS, problems of this nature are avoided by careful analysis of each active object's event handler, to ensure that event processing is kept as short as possible (described in Chapters 8 and 9). It would generally not be regarded as reason enough to multithread the code. However, there are occasions where the use of several threads may be necessary, say to perform a task which *cannot* be split into short-running increments. By implementing it in a separate thread, it can run asynchronously without impacting an application's response to user interface events.

On Symbian OS, active objects multitask cooperatively within a thread and cannot be pre-empted by the active scheduler; threads are scheduled pre-emptively by the kernel.

10.1 Class `RThread`

On Symbian OS, the class used to manipulate threads is `RThread` (you'll notice that it's an R class, the characteristics of which are described in Chapter 1). An object of type `RThread` represents a handle to a thread, because the thread itself is a kernel object. An `RThread` object can be used to create or refer to another thread in order to manipulate it (e.g. suspend, resume, panic or kill it) and to transfer data to or from it.

The `RThread` class has been modified quite significantly as part of the changes made for the new hard real-time kernel delivered in releases of Symbian OS v8.0. I'll identify the main differences as I come to them. Most notably, a number of the functions are now restricted for use on threads in the current process only, whereas previous versions of Symbian OS allowed one thread to manipulate any other thread in the system, even those in other processes. The changes to Symbian OS v8.0 have been introduced to protect threads against abuse from potentially malicious code.

At the time of going to press, Symbian identifies the hard real-time kernel as "EKA2" – which is a historical reference standing for "EPOC[2]

[2] Symbian OS was previously known as EPOC, and earlier still, EPOC32.

Kernel Architecture 2" – and refers to the kernel of previous releases as EKA1. Throughout this chapter, and the rest of the book, I'll use this nomenclature to distinguish between versions of Symbian OS v8.0 containing the new kernel and previous versions when discussing any API differences.

On Symbian OS v7.0, class RThread is defined as follows in e32std.h. I've included the entire class definition because I'll mention many of its methods throughout this chapter.

```
class RThread : public RHandleBase
    {
public:
    inline RThread();
    IMPORT_C TInt Create(const TDesC& aName, TThreadFunction aFunction,
            TInt aStackSize,TAny* aPtr,RLibrary* aLibrary,RHeap* aHeap,
            TInt aHeapMinSize,TInt aHeapMaxSize,TOwnerType aType);
    IMPORT_C TInt Create(const TDesC& aName, TThreadFunction aFunction,
            TInt aStackSize,TInt aHeapMinSize,TInt aHeapMaxSize,
            TAny* aPtr,TOwnerType aType=EOwnerProcess);
    IMPORT_C TInt Create(const TDesC& aName,TThreadFunction aFunction,
            TInt aStackSize,RHeap* aHeap,TAny* aPtr,
            TOwnerType aType=EOwnerProcess);
    IMPORT_C TInt SetInitialParameter(TAny* aPtr);
    IMPORT_C TInt Open(const TDesC& aFullName,
            TOwnerType aType=EOwnerProcess);
    IMPORT_C TInt Open(TThreadId aID,TOwnerType aType=EOwnerProcess);
    IMPORT_C TThreadId Id() const;
    IMPORT_C void Resume() const;
    IMPORT_C void Suspend() const;
    IMPORT_C TInt Rename(const TDesC& aName) const;
    IMPORT_C void Kill(TInt aReason);
    IMPORT_C void Terminate(TInt aReason);
    IMPORT_C void Panic(const TDesC& aCategory,TInt aReason);
    IMPORT_C TInt Process(RProcess& aProcess) const;
    IMPORT_C TThreadPriority Priority() const;
    IMPORT_C void SetPriority(TThreadPriority aPriority) const;
    IMPORT_C TProcessPriority ProcessPriority() const;
    IMPORT_C void SetProcessPriority(TProcessPriority aPriority) const;
    IMPORT_C TBool System() const;
    IMPORT_C void SetSystem(TBool aState) const;
    IMPORT_C TBool Protected() const;
    IMPORT_C void SetProtected(TBool aState) const;
    IMPORT_C TInt RequestCount() const;
    IMPORT_C TExitType ExitType() const;
    IMPORT_C TInt ExitReason() const;
    IMPORT_C TExitCategoryName ExitCategory() const;
    IMPORT_C void RequestComplete(TRequestStatus*& aStatus,
            TInt aReason) const;
    IMPORT_C TInt GetDesLength(const TAny* aPtr) const;
    IMPORT_C TInt GetDesMaxLength(const TAny* aPtr) const;
    IMPORT_C void ReadL(const TAny* aPtr,TDes8& aDes,
            TInt anOffset) const;
    IMPORT_C void ReadL(const TAny* aPtr,TDes16 &aDes,
            TInt anOffset) const;
    IMPORT_C void WriteL(const TAny* aPtr,const TDesC8& aDes,
            TInt anOffset) const;
```

```
    IMPORT_C void WriteL(const TAny* aPtr,const TDesC16& aDes,
            TInt anOffset) const;
    IMPORT_C void Logon(TRequestStatus& aStatus) const;
    IMPORT_C TInt LogonCancel(TRequestStatus& aStatus) const;
    IMPORT_C RHeap* Heap();
    IMPORT_C void HandleCount(TInt& aProcessHandleCount,
            TInt& aThreadHandleCount) const;
    IMPORT_C TExceptionHandler* ExceptionHandler() const;
    IMPORT_C TInt SetExceptionHandler(TExceptionHandler* aHandler,
            TUint32 aMask);
    IMPORT_C void ModifyExceptionMask(TUint32 aClearMask,
            TUint32 aSetMask);
    IMPORT_C TInt RaiseException(TExcType aType);
    IMPORT_C TBool IsExceptionHandled(TExcType aType);
    IMPORT_C void Context(TDes8& aDes) const;
    IMPORT_C TInt GetRamSizes(TInt& aHeapSize,TInt& aStackSize);
    IMPORT_C TInt GetCpuTime(TTimeIntervalMicroSeconds& aCpuTime)
            const;
    inline TInt Open(const TFindThread& aFind,
            TOwnerType aType=EOwnerProcess);
    };
```

The base class of RThread is RHandleBase, which encapsulates the behavior of a generic handle and is used as a base class throughout Symbian OS to identify a handle to another object, often a kernel object. As Chapter 1 discussed, CBase must always be the base class of a C class (if only indirectly), but RHandleBase is not necessarily always the base of an R class, although you'll find that a number of Symbian OS R classes do derive from it (e.g. RThread, RProcess, RMutex and RSessionBase). Neither does RHandleBase share the characteristics of class CBase (such as a virtual destructor and zero-initialization through an overloaded operator new). Instead, class RHandleBase encapsulates a 32-bit handle to the object its derived class represents and exports a limited number of public API methods to manipulate that handle, namely Handle(), SetHandle(), Duplicate() and Close().

You can use the default constructor of RThread to acquire a handle to the thread your code is currently running in, which can be used as follows:

```
RHeap* myHeap = RThread().Heap(); // Handle to the current thread's heap

// ...or...

_LIT(KClangerPanic, "ClangerThread");
// Panic the current thread with KErrNotFound
RThread().Panic(KClangerPanic, KErrNotFound);
```

The handle created by the default constructor is actually a pseudo-handle set to the constant KCurrentThreadHandle, which is treated specially by the kernel. If you want a "proper" handle to the current thread (that is, a handle that is in the thread handle list) to pass between threads in a process, you need to duplicate it using `RThread::Duplicate()`:

```
RThread properhandle.Duplicate(RThread());
```

You can acquire a handle to a different thread either by creating it or by opening a handle on a thread which currently exists in the system. As you can see from its definition, class `RThread` defines several functions for thread creation. Each function takes a descriptor representing a unique name for the new thread, a pointer to a function in which execution starts, a pointer to data to be passed to that function and a value for the stack size of the thread, which defaults to 8 KB. The `Create()` function is overloaded to allow you to set various options associated with the thread's heap, such as its maximum and minimum size and whether it shares the creating thread's heap or uses a specific heap. By default on Symbian OS, each thread has its own independent heap as well as its own stack. The size of the stack is limited to the size you set in `RThread::Create()`, but the heap can grow from its minimum size up to a maximum size, which is why both values may be specified in one of the overloads of `Create()`.[3] Where the thread has its own heap, the stack and the heap are located in the same chunk of memory.

When the thread is created, it is assigned a unique thread identity, which is returned by the `Id()` function of `RThread` as a `TThreadId` object. If you know the identity of an existing thread, you can pass it to `RThread::Open()` to open a handle to that thread. Alternatively, you can use an overload of the `Open()` function to pass the unique name of a thread on which to open a handle. The thread's name is set as a parameter of the `Create()` function overloads when the thread is created; once you have a handle to a thread, it is possible to rename it using `RThread::Rename()` if the thread is not protected.

`RThread` has been modified in releases of Symbian OS v8.0 to protect threads against abuse from potentially malicious code.

10.2 Thread Priorities

On Symbian OS, threads are pre-emptively scheduled and the currently running thread is the highest priority thread ready to run. If there are two

[3] You can also specify the minimum and maximum heap size for the main thread of a component that runs as a separate process in its `.mmp` file, using the following statement:

```
epocheapsize minSizeInBytes maxSizeInBytes
```

or more threads with equal priority, they are time-sliced on a round-robin basis. The priority of a thread is a number: the higher the value, the higher the priority.

When writing multithreaded code, you should consider the relative priorities of your threads carefully. You should not arbitrarily assign a thread a high priority, otherwise it may pre-empt other threads in the system and affect their response times. Even those threads which may legitimately be assigned high priorities must endeavor to make their event-handling service complete rapidly, so they can yield the system and allow other threads to execute.

A thread has an absolute priority which is calculated from the priority assigned to the thread by a call to `RThread::SetPriority()` and optionally combined with the priority assigned to the process in which it runs. The `TThreadPriority` and `TProcessPriority` enumerations, taken from `e32std.h`, are shown below.

```
enum TThreadPriority
    {
    EPriorityNull=(-30),
    EPriorityMuchLess=(-20),
    EPriorityLess=(-10),
    EPriorityNormal=0,
    EPriorityMore=10,
    EPriorityMuchMore=20,
    EPriorityRealTime=30,
    EPriorityAbsoluteVeryLow=100,
    EPriorityAbsoluteLow=200,
    EPriorityAbsoluteBackground=300,
    EPriorityAbsoluteForeground=400,
    EPriorityAbsoluteHigh=500
    };

enum TProcessPriority
    {
    EPriorityLow=150,
    EPriorityBackground=250,
    EPriorityForeground=350,
    EPriorityHigh=450,
    EPriorityWindowServer=650,
    EPriorityFileServer=750,
    EPriorityRealTimeServer=850,
    EPrioritySupervisor=950
    };
```

The following values are relative thread priorities: `EPriorityMuchLess`, `EPriorityLess`, `EPriorityNormal`, `EPriorityMore` and `EPriorityMuchMore`. If these values are passed to `RThread::SetPriority()`, the resulting absolute priority of the thread is the combination of the priority of the process and the relative value specified. For example, if the process has `TProcessPriority::EPriorityHigh` (=450), and `SetPriority()` is called with `TThreadPriority` of `EPriorityMuchLess`, the absolute priority of the thread will be

450 − 20 = 430. If, for the same process, TThreadPriority::EPriorityMuchMore is passed to SetPriority() instead, the absolute priority of the thread will be 450 + 20 = 470.

The remaining TThreadPriority values, except EPriorityRealTime,[4] are absolute priorities that allow the priority of the thread to be independent of the process in which it is running. If these values are passed to a call to RThread::SetPriority(), the priority of the thread is set to the value specified, and the process priority is ignored. For example, if the process has TProcessPriority::EPriorityHigh (=450), the absolute priority of the thread can range from EPriorityAbsoluteVeryLow (=100) to EPriorityAbsoluteVeryHigh (=500), depending on the absolute TThreadPriority value selected.

All threads are created with priority EPriorityNormal by default. When a thread is created it is initially put into a suspended state and does not begin to run until Resume() is called on its handle. This allows the priority of the thread to be changed by a call to SetPriority() before it starts to run, although the priority of a thread can also be changed at any time.

The secure RThread class in EKA2 does not allow you to call SetPriority() on a handle to any thread except that in which the code is currently running. This prevents a badly programmed or malicious thread from modifying the priorities of other threads in the system, which may have a serious effect on the overall system performance.

10.3 Stopping a Running Thread

A running thread can be removed from the scheduler's ready-to-run queue by a call to Suspend() on its thread handle. It still exists, however, and can be scheduled to run again by a call to Resume(). A thread can be ended permanently by a call to Kill() or Terminate(), both of which take an integer parameter representing the exit reason. You should call Kill() or Terminate() to stop a thread normally, reserving Panic() for stopping the thread to highlight a programming error. If the main thread in a process is ended by any of these methods, the process terminates too.

On EKA1, a thread must call SetProtected() to prevent other threads from acquiring a handle and stopping it by calling Suspend(), Panic(), Kill() or Terminate(). On EKA2, the security model ensures that a thread is always protected and the redundant SetProtected() method has been removed. This default protection ensures

[4] TThreadPriority::EPriorityRealTime is treated slightly differently. Passing this value to SetPriority() causes the thread priority to be set to TProcessPriority::EPriorityRealTimeServer.

that it is no longer possible for a thread to stop another thread in a separate process by calling `Suspend()`, `Terminate()`, `Kill()` or `Panic()` on it. The functions are retained in EKA2 because a thread can still call the various termination functions on itself or other threads in the same process. `RMessagePtr` has `Kill()`, `Terminate()` and `Panic()` methods that are identical to those in `RThread`. These allow a server to terminate a client thread, for example to highlight a programming error by causing a panic in a client which has passed invalid request data. Chapters 11 and 12 discuss the client–server framework in more detail.

The manner by which the thread was stopped, and its exit reason, can be determined from the `RThread` handle of an expired thread by calling `ExitType()`, `ExitReason()` and `ExitCategory()`. The exit type indicates whether `Kill()`, `Terminate()` or `Panic()` was called or, indeed, if the thread is still running. The exit reason is the integer parameter value passed to the `Kill()` or `Terminate()` functions, or the panic reason. The category is a descriptor containing the panic category, "`Kill`" or "`Terminate`". If the thread is still running, the exit reason is zero and the exit category is a blank string.

It is also possible to receive notification when a thread dies. A call to `RThread::Logon()` on a valid thread handle, passing in a `TRequestStatus` reference, submits a request for notification when that thread terminates. The request completes when the thread terminates and receives the value with which the thread ended or `KErrCancel` if the notification request was cancelled by a call to `RThread::LogonCancel()`.

The following example code, which is suitable both for EKA1 and EKA2, demonstrates how you may use thread termination notification. It shows how some long-running synchronous function (`SynchronousTask()`) can be encapsulated within an active object class which runs the task in a separate thread, allowing the function to be called asynchronously. This is useful if a caller doesn't want to be blocked on a slow-to-return synchronous function. For example, a third-party library may provide a synchronous API for preparing a file for printing. The UI code which uses it cannot just "hang" while waiting for the function to return; it requires the operation to be performed asynchronously. Of course, ideally, the function would be asynchronous, implemented incrementally in a low-priority active object, as described in Chapter 9. However, some tasks cannot easily be split into small units of work, or may be ported from code which was not designed for active objects.

The class defined below supplies an asynchronous wrapper, `DoAsyncTask()`, over a synchronous function, allowing the caller to submit an asynchronous request and receive notification of its completion through a `TRequestStatus` object.

```
_LIT(KThreadName, "ExampleThread"); // Name of the new thread

TInt SynchronousTask(); // Example of a long-running synchronous function

class CAsyncTask : public CActive
    {// Active object class to wrap a long synchronous task
public:
    ~CAsyncTask();
    static CAsyncTask* NewLC();
    // Asynchronous request function
    void DoAsyncTask(TRequestStatus& aStatus);
protected: // From base class
    virtual void DoCancel();
    virtual void RunL();
    virtual TInt RunError(TInt anError);
private:
    CAsyncTask();
    void ConstructL();
    // Thread start function
    static TInt ThreadEntryPoint(TAny* aParameters);
private:
    TRequestStatus* iCaller; // Caller's request status
    RThread iThread;         // Handle to created thread
    };
```

The implementation of the active object class is shown below. `DoAsyncTask()` is the asynchronous request-issuing function into which the caller passes a `TRequestStatus` object, typically from another active object, which is stored internally as `iCaller` by the function. Additionally, `DoAsyncTask()` creates the thread in which the synchronous function will run, calls `Logon()` upon the thread to receive notification when it terminates, then sets itself active before resuming the thread.

When the thread terminates, the `Logon()` request completes (the `iStatus` of the active object receives the exit reason). The active scheduler calls `RunL()` on the active object which, in turn, notifies the caller that the function call has completed by calling `User::Request-Complete()` on `iCaller`, the stored `TRequestStatus` object.

The class implements the `RunError()` and `DoCancel()` functions of the `CActive` base class. In `DoCancel()`, the code checks whether the thread is still running because a case could arise where the long-running function has completed and the thread has ended, but the resulting event notification has not yet been handled by the active scheduler. If the thread is outstanding, `DoCancel()` calls `Kill()` on it, closes the thread handle and completes the caller with `KErrCancel`.

```
CAsyncTask::CAsyncTask()
: CActive(EPriorityStandard) // Standard priority unless good reason
    {// Add to the active scheduler
    CActiveScheduler::Add(this);
    }
```

```
// Two-phase construction code omitted for clarity

CAsyncTask::~CAsyncTask()
    {// Cancel any outstanding request before cleanup
    Cancel(); // Calls DoCancel()

    // The following is called by DoCancel() or RunL()
    // so is unnecessary here too
    // iThread.Close(); // Closes the handle on the thread
    }

void CAsyncTask::DoAsyncTask(TRequestStatus& aStatus)
    {
    if (IsActive())
        {
        TRequestStatus* status = &aStatus;
        User::RequestComplete(status, KErrAlreadyExists);
        return;
        }

    // Save the caller's TRequestStatus to notify them later
    iCaller = &aStatus;

    // Create a new thread, passing the thread function and stack sizes
    // No extra parameters are required in this example, so pass in NULL
    TInt res = iThread.Create(KThreadName, ThreadEntryPoint,
            KDefaultStackSize, NULL, NULL);
    if (KErrNone!=res)
        {// Complete the caller immediately
        User::RequestComplete(iCaller, res);
        }
    else
        {// Set active; resume new thread to make the synchronous call
         // (Change the priority of the thread here if required)

         // Set the caller and ourselves to KRequestPending
         // so the active scheduler notifies on completion
        *iCaller = KRequestPending;
        iStatus = KRequestPending;
        SetActive();
        iThread.Logon(iStatus); // Request notification when thread dies
        iThread.Resume();       // Start the thread
        }
    }

TInt CAsyncTask::ThreadEntryPoint(TAny* /*aParameters*/)
    {// Perform a long synchronous task e.g. a lengthy calculation
    TInt res = SynchronousTask();
    // Task is complete so end this thread with returned error code
    RThread().Kill(res);
    return (KErrNone);    // This value is discarded
    }

void CAsyncTask::DoCancel()
    {// Kill the thread and complete with KErrCancel
     // ONLY if it is still running
    TExitType threadExitType = iThread.ExitType();
    if (EExitPending==threadExitType)
```

```
        {// Thread is still running
        iThread.LogonCancel();
        iThread.Kill(KErrCancel);
        iThread.Close();
        // Complete the caller
        User::RequestComplete(iCaller, KErrCancel);
        }
    }

void CAsyncTask::RunL()
    {// Check in case thread is still running e.g. if Logon() failed
    TExitType threadExitType = iThread.ExitType();
    if (EExitPending==threadExitType) // Thread is still running, kill it
        iThread.Kill(KErrNone);

    // Complete the caller, passing iStatus value to RThread::Kill()
    User::RequestComplete(iCaller, iStatus.Int());
    iThread.Close(); // Close the thread handle, no need to LogonCancel()
    }

TInt CAsyncTask::RunError(TInt anError)
    {
    if (iCaller)
        {
        User::RequestComplete(iCaller, anError);
        }
    return (KErrNone);
    }
```

For more sophisticated thread death notification, you can alternatively use the `RUndertaker` thread-death notifier class, which is described in detail in the SDK. By creating an `RUndertaker`, you receive notification when any thread is about to die. The `RUndertaker` passes back a notification for each thread death, including the exit reason and the `TThreadId`. It creates a thread-relative handle to the dying thread, which effectively keeps it open. This means that you must close the thread handle to release the thread finally into the abyss – hence the name of the class. This notification is useful if you want to track the death of every thread in the system, but if you're interested in the death of one specific thread, it's easier to use `RThread::Logon()`, as described above.

It is possible to receive notification when a thread dies by making a call to `RThread::Logon()` on a valid thread handle. The manner and reason for thread termination can also be determined from the `RThread` handle of an expired thread by calling `ExitType()`, `ExitReason()` and `ExitCategory()`.

10.4 Inter-Thread Data Transfer

On Symbian OS, you can't transfer data pointers directly between threads running in separate processes, because process address spaces are protected from each other, as I described in Chapter 8.[5]

On EKA1 versions of Symbian OS, RThread provides a set of functions that enable inter-thread data transfer regardless of whether the threads are in the same or different processes:

```
IMPORT_C TInt GetDesLength(const TAny* aPtr) const;
IMPORT_C TInt GetDesMaxLength(const TAny* aPtr) const;
IMPORT_C void ReadL(const TAny* aPtr,TDes8& aDes,TInt anOffset) const;
IMPORT_C void ReadL(const TAny* aPtr,TDes16 &aDes,TInt anOffset) const;
IMPORT_C void WriteL(const TAny* aPtr,const TDesC8& aDes,TInt anOffset)
      const;
IMPORT_C void WriteL(const TAny* aPtr,const TDesC16& aDes,TInt anOffset)
      const;
```

On EKA2 these methods have been withdrawn from class RThread, but are implemented by RMessagePtr, which can be used by a server to transfer data to and from a client thread. The functions take slightly different parameters but perform a similar role. I'll discuss inter-thread data transfer here using the RThread API, which is applicable to EKA1 releases. In EKA2 the RMessagePtr API is more intuitive, using a parameter value rather than TAny* to identify the source or target descriptor in the "other" thread. You should consult an appropriate v8.0 SDK for complete documentation.

On EKA1, RThread::WriteL() can be used to transfer data from the currently running thread to the thread represented by a valid RThread handle. RThread::WriteL() takes a descriptor (8- or 16-bit) containing the data in the current thread and writes it to the thread on whose handle the function is called. The destination of the data in that other thread is also a descriptor, which should be modifiable (derived from TDes). It is identified as a const TAny pointer into the other thread's address space. The function leaves with KErrBadDescriptor if the TAny pointer does not appear to point to a valid descriptor. The maximum length of the target descriptor can be determined before writing by a call to RThread::GetDesMaxLength().

[5] This is true for EKA2 and previous releases of Symbian OS running on target hardware. However, the Symbian OS Windows emulator for releases prior to EKA2 does not protect the address spaces of Symbian OS processes. The threads share writable memory, which means that each emulated process can be accessed by other processes. Code which uses direct pointer access to transfer data between threads in different processes will appear to work on the Windows emulator, then fail spectacularly when deployed on a real Symbian OS handset.

RThread::ReadL() can be used to transfer data into the currently running thread from the thread represented by a valid RThread handle. ReadL() transfers data from a descriptor in the "other" thread, identified as a const TAny pointer, into a modifiable descriptor (8- or 16-bit) in the current thread. The function leaves with KErrBadDescriptor if the TAny pointer does not point to what appears to be a valid descriptor. The length of the target descriptor can be determined before reading from it by a call to RThread::GetDesLength().

Inter-thread data transfer uses descriptors because they fully describe their length and maximum allowable length, which means that no extra parameters need to be transferred to indicate this information. If you want to transfer non-descriptor data, you can use the package classes TPckgC and TPckg, described in Chapter 6, which effectively "descriptorize" the data. Package classes are only valid for transferring a flat data structure such as that found in a T class object; their use in inter-thread data transfer between client and server threads is discussed in Chapter 12.

10.5 Exception Handling

On EKA1, the RThread API supports thread exception management. On the more secure EKA2 platform, this has moved into class User and applies only to the current thread. The signatures of the five functions are the same, however, regardless of the class that implements them.

```
TExceptionHandler* ExceptionHandler() const;
TInt SetExceptionHandler(TExceptionHandler* aHandler, TUint32 aMask);
void ModifyExceptionMask(TUint32 aClearMask, TUint32 aSetMask);
TInt RaiseException(TExcType aType);
TBool IsExceptionHandled(TExcType aType);
```

SetExceptionHandler() allows you to define an exception handler function for the thread for which the handle is valid. It takes an exception-handling function and a bitmask that allows you to specify the category of exception which will be handled (see e32std.h).

```
const TUint KExceptionAbort=0x01;
const TUint KExceptionKill=0x02;
const TUint KExceptionUserInterrupt=0x04;
const TUint KExceptionFpe=0x08;
const TUint KExceptionFault=0x10;
const TUint KExceptionInteger=0x20;
const TUint KExceptionDebug=0x40;
```

If an exception is raised on the thread that falls into the category handled, it will be passed to the specified exception handler to be dealt with.

The exception handler function, `TExceptionHandler`, is a `typedef` for a function which takes a value of `TExcType` (an enumeration which further identifies the type of exception) and returns `void`. Each of the `TExcType` enumeration values each correspond to one of the categories above but they identify the exception further, for example `EExcBounds-Check`, `EExcInvalidOpCode` and `EExcStackFault` are all types of the `KExceptionFault` category.

What should an exception handler function do? Well, you cannot identify where the problem occurred from within the handler, so the only thing you can do is leave from the handler function. Effectively, the handler converts an exception into a leave which can be caught and dealt with. However, by doing this, you require a `TRAP` around every operation that could generate an exception of that type. This could have a significant impact in terms of code size and speed (as discussed in Chapter 2) and as a result, while this technique might be useful for debugging, it is not recommended for code you intend to release.

Without a handler specified, the exception terminates the thread with a `KERN-EXEC 3` panic if it occurs during a kernel executive call, or a `USER-EXEC 3` panic if it occurs in a non-system call.

10.6 Processes

I introduced processes in Chapter 8 and briefly mentioned them in this chapter when discussing relative thread priorities. The `RProcess` class can be used to get a handle to a running process, in much the same way as `RThread` for a running thread. `RProcess` is defined in `e32std.h` and a number of the functions look quite similar to those in `RThread`.[6]

You can use the default constructor to create a handle to the current process, the `Create()` function to start a new, named process and the `Open()` function to open a handle to a process identified by name or process identity (`TProcessId`). The `Resume()` functions and assorted functions to stop the process will also look familiar. Note that there is no `Suspend()` function because processes are not scheduled; threads form the basic unit of execution and run inside the protected address space of a process.

Chapter 13 discusses the difference between the emulator and target hardware in more detail, but one aspect is particularly relevant here. On Windows, the emulator runs within a single Win32 process, `EPOC.exe`, and each Symbian OS process runs as a separate thread inside it.

[6] The SDK documentation for `RThread` and `RProcess` is extensive and should be consulted for further details of both these classes. Like the `RThread` class, `RProcess` has been secured in EKA2, resulting in the removal of some functions and the restriction of others to apply only to the currently running process

On EKA1 the emulation of processes on Windows is incomplete and `RProcess::Create()` returns `KErrNotFound`. To emulate any code that spawns a separate process, e.g. server startup code, different source code is required for emulator builds, using threads instead of processes in which to launch the server. The EKA2 release has removed this inconvenience. While Symbian OS still runs in a single process on Windows (indeed, that's what the "`S`" in `WINS`, `WINSCW` and `WINSB` stands for), the emulation is enhanced and you can use the same code, calling `RProcess::Create()` on both platforms. Symbian OS, for emulator builds, translates this call to creation of a new Win32 thread within `EPOC.exe`.

I'll conclude this chapter with a brief description of the most fundamental process within Symbian OS, the kernel server process, and the threads which run within it. I'll restrict the discussion to Symbian OS EKA1 – you will be able to find more detail of the changes introduced by EKA2 from the relevant system documentation (***www.symbian.com/technology/product_descriptions.html***). On EKA1, the kernel server process is a special process that contains two threads which run at supervisor privilege level:

- the kernel server thread
  This is the first thread to run in the system and has the highest priority of all threads. It implements all kernel functions requiring allocation or deallocation on the kernel heap and manages hardware resources. Device drivers run in the kernel thread and use interrupt service routines and delayed function calls to service events from the hardware.

- the "Null" or "Idle" thread
  This has the lowest priority in the system and therefore runs when no other threads are ready to run. It switches the processor into idle mode to save power, by calling a callback function to shut down part of the hardware. This allows Symbian OS to conserve power most efficiently, thus maximizing the battery life. To allow the Null thread to run, your code should wait on events (typically using active objects or kernel signaling primitives such as mutexes) rather than polling constantly. A polling thread prevents the Null thread from running and stops Symbian OS from entering its idle, power-saving, mode.

The other threads in the system are known as user threads and run in an unprivileged state. User threads can access kernel services, such as other threads or processes, using the APIs in the user library (`EUser.dll`). When a user thread requires a kernel service, `EUser` either switches to the kernel executive, which runs kernel-privileged code in the context of the running thread, or makes a call via the executive into the kernel server process.

10.7 Summary

This chapter compared the use of pre-emptively scheduled threads on Symbian OS with cooperatively scheduled active objects. Active objects are preferred for most multitasking code because they have lower runtime overheads. However, threads are sometimes necessary, for example if priority-based pre-emption is required or to perform a long-running background task asynchronously where it cannot be split into increments.

The chapter reviewed the `RThread` class, which represents a handle to a thread on Symbian OS. It identified the differences in the `RThread` API between EKA1 and EKA2 releases, because EKA2 has removed the facility for one thread to affect any other thread running in a separate process. The `RThread` API allows a thread handle to rename a thread, modify its priority and suspend, terminate or panic it. It also provides a means by which data may be transferred between threads (inter-thread data transfer is discussed in more detail in Chapters 11 and 12).

The chapter concluded by discussing Symbian OS processes in brief, and included a short description of the threads running with the EKA1 kernel server process.

The use of active objects for multitasking on Symbian OS is discussed in Chapters 8 and 9.

11

The Client–Server Framework in Theory

In theory there is no difference between theory and practice. In practice there is

Yogi Berra

This chapter examines the client–server framework model on Symbian OS. I'll be discussing the theory, such as why client–server is used on Symbian OS, how it works and the implementation classes involved. There's a lot of information, so I'll use a question and answer structure to break the chapter down clearly into logical sections.

When programming for Symbian OS, it is likely that you will often use system servers transparently, via classes that wrap the actual details of client–server communication. It's less common to implement your own server or write client-side code that accesses a server directly. So in this chapter, I'll make it clear which information you can skip if you really only want to know the basics of the Symbian OS client–server architecture and how to use a server as a typical client. I'll illustrate the chapter with examples of how to use a typical Symbian OS system server, the file server.

The next chapter works through some example code for a Symbian OS client and server, to illustrate how a typical implementation is put together and explain how to go about writing your own.

This chapter discusses the client–server model for Symbian OS releases up to and including v7.0s (the code samples use the client–server APIs from Symbian OS v7.0). On Symbian OS v8.0, the concepts are generally the same, but some of the APIs have changed. Rather than confuse matters, I've decided to concentrate solely on the releases available at the time of going to press. You can find more information about the differences in v8.0 from the relevant v8.0 SDKs when they become available, from other Symbian press books and, I hope, later editions of this book, or from the Symbian website (***www.symbian.com/technology.html***).

11.1 Why Have a Client–Server Framework?

A client makes use of services provided by a server. The server receives request messages from its clients and handles them, either synchronously or asynchronously.[1] Data is passed from the client to the server in the request message itself or by passing a pointer to a descriptor in the client address space, which the server accesses using kernel-mediated data transfer.

On Symbian OS, servers are typically used to manage shared access to system resources and services. The use of a server is efficient since it can service multiple client sessions and can be accessed concurrently by clients running in separate threads.

A server also protects the integrity of the system because it can ensure that resources are shared properly between clients, and that all clients use the resources correctly. When the server runs in its own process it has a separate, isolated address space and the only access a client has to the services in question is through a well-defined interface. By employing a server in a separate process, the system can guarantee that misbehaved clients cannot corrupt any of the resources the server manages on its behalf or for other clients. On Symbian OS, it is this rationale which protects the filesystem; all access to file-based data must be made through the file server client, `efsrv.dll`, or a higher-level component which uses it such as the stream store, `estor.dll`.

Servers can also be used to provide asynchronous services because they run in a separate thread to their clients. Most of the system services on Symbian OS, particularly those providing asynchronous functionality, are provided using the client–server framework: for example the window server (for access to UI resources such as the screen and keypad), the serial communications server (for access to the serial ports) and the telephony server.

11.2 How Do the Client and Server Fit Together?

A Symbian OS server always runs in a separate thread to its clients and often runs in a separate process. All interaction is performed by message passing or inter-thread data transfer.

[1] A **synchronous** function performs a service then returns directly to the caller, often returning an indication of success or failure. A typical example of a synchronous function is a call to format a descriptor or to copy the contents of one descriptor into another.

An **asynchronous** function submits a request as part of the function call and returns to the caller, but completion of that request occurs some time later. Upon completion, the caller receives a signal to notify it and indicate success or failure. A good example of an asynchronous service is a timer – the request completes at a later stage when the wait time has elapsed.

This protects the system resource that the server accesses from any badly programmed or malicious clients. A badly-behaved client should not be able to "crash" a server, although, of course, the server must still guard against invalid data or out-of-sequence client requests and, if necessary, panic the client responsible.

A typical server has associated client-side code that formats requests to pass to the server, via the kernel, and hides the implementation details of the private client–server communication protocol.

This means that, for example, a "client" of the Symbian OS file server (`efile.exe`[2]) is actually a client of the file server's client-side implementation and links against the DLL which provides it (`efsrv.dll`), as shown in Figure 11.1.

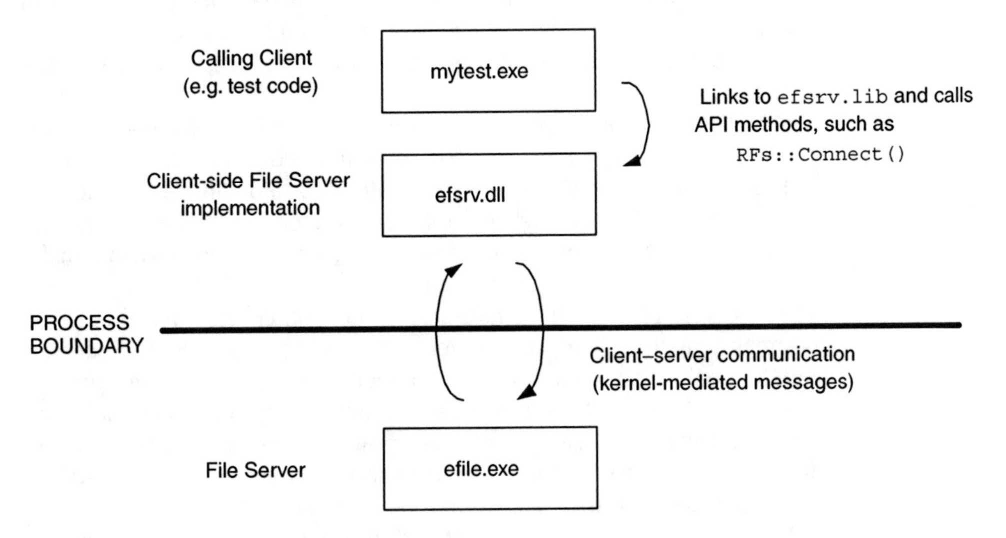

Figure 11.1 File server, client and calling client

In the rest of this chapter, I'll make it clear when I use the term "client" whether I mean:

- the client-side implementation that communicates directly with the server, or
- code which links to that client-side implementation, using it to request a service of the server, a "calling client".

[2] Chapter 13 discusses the difference between executable code which runs on the emulator and on target hardware. In effect, on Windows, Symbian OS runs in a single process which emulates multiple processes by loading them as DLLs into separate threads. Thus, while the file server is built to run on hardware as `efile.exe`, to run on Windows builds it is built as `efile.dll`.

The client-side server implementation may also be responsible for starting the server if it is not critical to the system (whereupon it will be started when the operating system itself starts).

11.3 How Do the Client and Server Communicate?

A client and a server run in different threads and often in different processes. When running in different processes, they cannot access each other's virtual address spaces, so they use a message-passing protocol to communicate. The communication channel is known as a **session**. A session is created by the kernel, which also acts as an intermediary for all client–server communication. A client may have several sessions with a server, although, as I'll explain later, the number of sessions should be kept to a minimum, because each session consumes resources in the kernel and server.

Client–server communication occurs when the client makes a request to the server using a message that identifies the nature of the request and can additionally hold some parameter data. For simple transactions this is sufficient, but for more complex parameters the server can transfer additional data to or from the client thread using inter-thread data transfer functions, which I described in Chapter 10.

In the rest of this chapter, unless I state otherwise, this discussion assumes that the client and server are running in separate processes, which means that data transfer between them requires inter-process communication (IPC). Under these circumstances parameter data can never be transferred using simple C++ pointers, because the server never has direct access to the client's address space (or vice versa). Instead, the kernel performs the data access on the server's behalf.

When the request has been fulfilled, the server notifies the client that it has completed by signaling the client thread's request semaphore, returning a completion result as a 32-bit value.

11.4 What Classes Does the Client–Server Framework Use?

This section gives an overview of the classes Symbian OS uses to implement the client–server framework. For further details of any of the classes, you should consult your preferred SDK. The next chapter reviews a typical implementation of a **transient** server[3] and its client-side access code, and further illustrates how the classes I discuss here are used.

---

[3] A transient server is started by the first client session that connects to it and shuts itself down when the last client session disconnects itself. I'll discuss server startup and shutdown in more detail in Section 11.6.

This section provides general background information on how the client–server framework is implemented. Although it helps to understand what is happening "under the hood", this information is not necessary simply to use a Symbian OS server. Most of the rest of the chapter discusses the implementation of a typical server or the client-side wrapper code that communicates with a server. If you don't want to get into this much detail, but want to know how to use a Symbian OS server most effectively, you may wish to skip ahead to Section 11.14.

The classes I'll discuss are as follows:

- `RSessionBase` – the client-side base class, representing a session with a server
- `RMessage` – a server-side representation of a client request and its request payload data
- `DSession` – a kernel class that represents a client–server session
- `CSharableSession` – an abstract base class for a server-side representation of a session
- `CServer` – an abstract base class, deriving from `CActive`, which is used server-side to receive client requests from the kernel and direct them to the appropriate session.

RSessionBase

`RSessionBase` is the main client-side class. It derives from `RHandleBase`, which is the base class for classes that own handles to other objects, often those created within the kernel. `RSessionBase` uniquely identifies a client–server session.

Here's the declaration of `RSessionBase` from `e32std.h` (I've omitted a couple of private methods):

```
class RSessionBase : public RHandleBase
    {
public:
    enum TAttachMode {EExplicitAttach,EAutoAttach};
public:
    IMPORT_C TInt Share(TAttachMode aAttachMode=EExplicitAttach);
    IMPORT_C TInt Attach() const;
protected:
    inline TInt CreateSession(const TDesC& aServer,
            const TVersion& aVersion);
    IMPORT_C TInt CreateSession(const TDesC& aServer,
            const TVersion& aVersion,TInt aAsyncMessageSlots);
    IMPORT_C TInt Send(TInt aFunction,TAny* aPtr) const;
    IMPORT_C void SendReceive(TInt aFunction,TAny* aPtr,
            TRequestStatus& aStatus) const;
    IMPORT_C TInt SendReceive(TInt aFunction,TAny* aPtr) const;
    };
```

The methods of this class are used to send messages to the server. You'll notice that most of them are protected. This is because the client-side class which accesses a server will typically derive from RSessionBase (for example, class RFs, which provides access to the file server). The derived class exports functions that wrap RSessionBase communication with the server and are more meaningful to potential clients of the server (such as RFs::Delete() or RFs::GetDir()).

The overloads of RSessionBase::CreateSession() start a new client–server session. They are typically called by client-side implementation code in an exported method such as Open() or Connect(). As an example, when you start a session with the file server you call RFs::Connect(), which itself calls RSessionBase::CreateSession(). When the session is opened successfully, corresponding kernel and server-side objects are created.

A server has a unique name which must be passed to RSessionBase::CreateSession() to connect the client to the correct server. Again, the client-side implementation takes care of this, so the calling client does not need to know the name of the server. CreateSession() also takes a TVersion structure[4] for compatibility support.

You'll notice that one overload of CreateSession() takes an integer parameter called aAsyncMessageSlots. This value reserves a number of slots to hold any outstanding asynchronous requests that client session may have with the server.[5] The maximum number of slots that may be reserved for each server is 255. The other overload of CreateSession() does not pre-allocate a maximum number of message slots. Instead, they are taken from a kernel-managed pool, of up to 255 message slots for that server, which is available to the whole system. If the number of outstanding requests to a server exceeds the number of slots in the system pool, or the number reserved for a particular session, the asynchronous request fails to be submitted and completes immediately with the error KErrServerBusy.

A request to a server is issued through a call to RSessionBase::SendReceive() or RSessionBase::Send(). SendReceive() has overloads to handle both synchronous and asynchronous requests. The asynchronous request method takes a TRequestStatus& parameter, while the synchronous version returns the result in the TInt return value. RSessionBase::Send() sends a message to the server but does not receive a reply (and in practice, this function is rarely used).

The Send() and SendReceive() methods take a 32-bit argument (aFunction) that identifies the client request (typically defined in an enumeration shared between the client and server – see the THerculeanLabors enumeration in the sample code of Chapter 12

[4] A TVersion object contains three integers representing the major, minor and build version numbers.

[5] A session can only ever have one outstanding synchronous request with a server.

for an example). The methods also take a `TAny*` argument which is a pointer to an array of four 32-bit values. This array constitutes the "payload" for the request; it can be empty or it may contain up to four integer values or pointers to descriptors in the client's address space. The layout of the array is determined in advance for each request between client and server and is a private protocol. The calling client does not need to have any knowledge of how data is transferred.

If a server supports session sharing, a client session may be shared by all the threads in a client process (Symbian OS v8.0 is the first release which also allows a session to be shared between processes). However, some servers restrict the session to the thread which connected to the server alone (this was always the case until sharable sessions were introduced in Symbian OS v6.0). I'll discuss sharable sessions in more detail later in this chapter.

On the client side, if a session can be shared, the first connection to the server should be made as normal using `RSessionBase::CreateSession()`. Once the session is opened, `RSessionBase::Share()` should be called on it to make it sharable.[6] Until `Share()` is called, the session is specific to the connecting client thread. If the `TAttachMode` parameter passed to `Share()` is `EExplicitAttach`, other threads wishing to share the session should call `RSessionBase::Attach()` on the session. However, if `EAutoAttach` is passed to `Share()`, then all threads are attached to the session automatically. If the session is not attached before a message is sent, or the session is not sharable, a panic occurs (`KERN-SVR` 0).

Client requests are identified using an enumeration shared between the client and server. Requests are submitted with a payload array which can contain up to four 32-bit values (integer values or pointers to descriptors in the client's address space).

RMessage

An `RMessage` object is a server-side representation of a client request, and each client request to the server is represented by a separate `RMessage` object. Here is the definition from `e32std.h`, where again, I've shown only the most relevant methods:

```
class RMessage
    {
public:
    IMPORT_C RMessage();
```

[6] A server must support sharable sessions, otherwise a call to `RSessionBase::Share()` raises a panic (`KERN-SVR` 23).

```
    IMPORT_C void Complete(TInt aReason) const;
    IMPORT_C void ReadL(const TAny* aPtr,TDes8& aDes) const;
    IMPORT_C void ReadL(const TAny* aPtr,TDes8& aDes,
           TInt anOffset) const;
    IMPORT_C void ReadL(const TAny* aPtr,TDes16& aDes) const;
    IMPORT_C void ReadL(const TAny* aPtr,TDes16& aDes,
           TInt anOffset) const;
    IMPORT_C void WriteL(const TAny* aPtr,const TDesC8& aDes) const;
    IMPORT_C void WriteL(const TAny* aPtr,const TDesC8& aDes,
           TInt anOffset) const;
    IMPORT_C void WriteL(const TAny* aPtr,const TDesC16& aDes) const;
    IMPORT_C void WriteL(const TAny* aPtr,const TDesC16& aDes,
           TInt anOffset) const;
    IMPORT_C void Panic(const TDesC& aCategory,TInt aReason) const;
    IMPORT_C void Kill(TInt aReason) const;
    IMPORT_C void Terminate(TInt aReason) const;
    inline TInt Function() const;
    inline const RThread& Client() const;
    inline TInt Int0() const;
    inline TInt Int1() const;
    inline TInt Int2() const;
    inline TInt Int3() const;
    inline const TAny* Ptr0() const;
    inline const TAny* Ptr1() const;
    inline const TAny* Ptr2() const;
    inline const TAny* Ptr3() const;
    inline const RMessagePtr MessagePtr() const;
protected:
    TInt iFunction;
    TInt iArgs[KMaxMessageArguments];
    RThread iClient;
    const TAny* iSessionPtr;
    TInt iHandle;
    };
```

The `RMessage` object stores the 32-bit request identifier (also known as an "opcode"), which can be retrieved by calling `Function()`. It also holds the array of request payload data, a handle to the client thread, accessible through `Client()`, and the client's `RHandleBase` identification handle.

As I described above, the layout of the request parameters in the request data array is pre-determined for each client request. For requests where the server expects parameters from the client, it can retrieve the data from an `RMessage` object using `Int0()` to return a 32-bit value from the first element of the request array, `Int1()` to return the second element, and so on. In a similar manner, `Ptr0()` returns the contents of the first element in the request array as a `TAny*` pointer, `Ptr1()` for the second element, and so on to the fourth element of the array.

The pointers returned from `Ptr0()` to `Ptr3()` *cannot* be used directly by the server code if they refer to the address space of a client running in a different process. The server must instead pass these pointers to the overloaded `ReadL()` and `WriteL()` methods[7] of `RMessage`, which use kernel-mediated inter-process communication to transfer the data.

When the server has serviced a client request, it calls `Complete()` on the `RMessage` to notify the client. This method wraps a call to `RThread::Complete()` on the client's thread handle. The integer value passed to `Complete()` is written into the client's `TRequestStatus` value and the request semaphore for the client thread is signaled. If you're wondering about synchronous `SendReceive()` requests from the client, which don't take a `TRequestStatus` parameter, take a look at Section 11.5. The `Panic()`, `Terminate()` and `Kill()` methods of `RMessage` are wrappers over the `RThread` methods of the same name and may be used by the server to stop the client thread under certain circumstances, such as client misbehavior due to a programming error.

The client and server run in separate threads which also typically run in different processes. The address spaces of Symbian OS processes are protected and kernel-mediated data transfer must be used between the client and server.

DSession

`DSession` is a kernel class; the "D" prefix means that it is a `CBase`-derived kernel-side class.

On Symbian OS, each process is memory-mapped into a different address space, so it is protected from other processes. One process cannot overwrite the memory of another. The only process that can "see" all the physical memory in the system is the kernel process. When a client calls `RSessionBase::CreateSession()`, the kernel establishes the connection between the client and the server and creates a `DSession` object to represent the session. Each `DSession` object has a pointer to a kernel object representing the client thread (`DThread`) and a pointer to the kernel server object (`DServer`), as shown in Figure 11.2.

CSharableSession

`CSharableSession` is an abstract base class that represents a session within the server. For each `RSessionBase`-derived object on the client side, there is an associated `CSharableSession`-derived object on the server side.

---

[7] These methods are simply wrappers over the `ReadL()` and `WriteL()` methods of `RThread`, as described in Chapter 10.

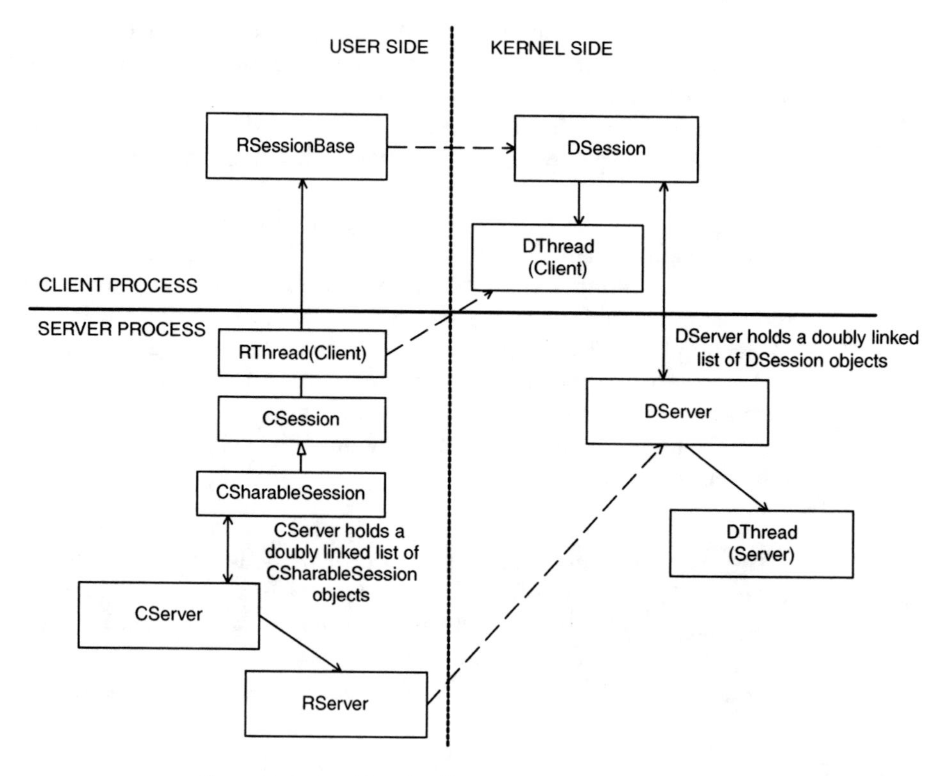

Figure 11.2 Symbian OS client- and server-side base classes

```
class CSharableSession : public CBase
    {
    friend class CServer;
public:
    IMPORT_C ~CSharableSession()=0;
    IMPORT_C virtual void CreateL(const CServer& aServer);
    inline const CServer* Server() const;
    inline const RMessage& Message() const;
    IMPORT_C void ResourceCountMarkStart();
    IMPORT_C void ResourceCountMarkEnd();
    IMPORT_C virtual TInt CountResources();
    virtual void ServiceL(const RMessage& aMessage)=0;
protected:
    IMPORT_C CSharableSession();
    ...
private:
    TInt iResourceCountMark;
    TDblQueLink iLink;
    const CServer* iServer;
    };
```

`CSharableSession` provides methods to access the `CServer`-derived object, `Server()`, which I'll discuss shortly. `Message()` can be used to access the next client request to process, if there any are outstanding. If the server makes an asynchronous call to service the request and does not complete the request before returning from `ServiceL()`, the `RMessage` object must be stored so it can be completed at a later stage. If it was not stored, when it came to complete the request, `Message()` would return a different message if other requests had been submitted to the server while the asynchronous request was being processed.

Classes derived from `CSharableSession` handle incoming client requests through their implementation of the pure virtual `ServiceL()` method. Typically, this method should check the incoming message to see which request the client has submitted, then handle it by unpacking the message and using the incoming parameters accordingly. When the request has been handled, the server calls `RMessage::Complete()` to notify the client thread of request completion. The example code in Chapter 12 illustrates this.

You'll notice a set of resource-counting functions, `ResourceCountMarkStart()`, `ResourceCountMarkEnd()` and `CountResources()`, which have a default, "do nothing" implementation in the `CSharableSession` base class but which may be overridden by derived classes for customized resource checking at server startup and shutdown, usually used only in debug builds. `ResourceCountMarkStart()` initializes server resource counting while `ResourceCountMarkEnd()` checks that the current number of server resources (e.g. subsessions) in use is equivalent to that when resource counting started. If the values are not equal, the function panics the client thread associated with the most recent message. `CountResources()` returns the number of server resources currently in use.

Prior to Symbian OS v6.0, class `CSession` represented a session on the server side. The `CSession` class was thread-specific and accessible only by a single thread on the client side. Symbian OS v6.0 introduced the concept of sharable client–server sessions. From v6.0, a client session may potentially be shared between multiple threads in the same client process, although a server implementation is not required to support sharable sessions. To support this modification, Symbian OS v6.0 introduced `CSharableSession` as the base class for a server-side session. `CSession` still exists, deriving from `CSharableSession`, as an abstract class which provides a set of thread-specific functions to transfer data between client and server.

CServer

The fundamental server-side base class is CServer, which itself derives from CActive. Here's the definition of CServer from e32base.h, with the less relevant details omitted for clarity:

```
class CServer : public CActive
    {
public:
    IMPORT_C ~CServer()=0;
    IMPORT_C TInt Start(const TDesC& aName);
    IMPORT_C void StartL(const TDesC& aName);
    IMPORT_C void ReStart();
    inline const RMessage& Message() const;
protected:
    IMPORT_C CServer(TInt aPriority,
            TServerType aType=EUnsharableSessions);
    IMPORT_C void DoCancel();
    IMPORT_C void RunL();
private:
    virtual CSharableSession* NewSessionL(const TVersion& aVersion)
            const=0;
    void Connect();
private:
    const TServerType iSessionType;
    RServer iServer;
    TDblQue<CSharableSession> iSessionQ;
protected:
    // to iterate the list of connected sessions
    TDblQueIter<CSharableSession> iSessionIter;
    };
```

The system ensures that there is only one CServer-derived active object created for each uniquely-named server. This object receives requests from all clients of the server as events, receiving notification of each incoming request from the kernel. By inspecting the RMessage associated with the request, the event handler method, CServer::RunL(), determines whether the CServer-derived object handles the requests itself or directs them to be handled by the appropriate server-side session class.

If the request is to connect a new session, CServer::RunL() calls the NewSessionL() method of the derived class, which creates a new server-side session object. If it is a request from a client-side session to disconnect, the associated server-side session object is destroyed in CServer::RunL(). For other client requests, CServer::RunL() calls the ServiceL() method of the associated CSharableSession-derived object to service the request. Having serviced each request, CServer::RunL() resubmits a "message receive" request and awaits further client requests.

When you are implementing a server, you must create an active scheduler as part of server startup (as I described in Chapter 8, which covers the basics of active objects). I'll illustrate how to do this in the example code for a typical server in the next chapter. `CServer::StartL()` adds the server to the active scheduler and initiates the first message receive request.

For each `CServer` object created in the system, a corresponding `DServer` object is created in the kernel. Each `DServer` object holds a doubly-linked queue of all `DSessions`, representing all the currently open sessions for that server. It also owns a kernel object, `DThread`, which represents the server thread.

When implementing a server, you must create an active scheduler during server startup.

11.5 How Do Synchronous and Asynchronous Requests Differ?

A client can request synchronous or asynchronous services[8] from a server. Asynchronous requests take a `TRequestStatus` reference parameter, which is passed to `RSessionBase::SendReceive()`. This parameter is filled with a completion result by the server, via the kernel, when the request is completed by a call to `RThread::RequestComplete()`, which also signals the request semaphore of the client thread to notify it of request completion.

In fact, synchronous requests to the server are actually "pseudo-synchronous". The synchronous overload of `RSessionBase::SendReceive()` declares a `TRequestStatus` object locally, passes this to the asynchronous overload of `SendReceive()` and then blocks the client thread until the request completes. In effect, the client thread is suspended and notified only when a server has completed its action, rather than continuing to poll the server for the status of a request. This is important on Symbian OS, to minimize power consumption.

11.6 How Is a Server Started?

There are several ways in which a server can be started and stopped:

- System servers, e.g. the file server, are started by Symbian OS as part of OS startup because they are essential to the operating system.

[8] For each asynchronous request function a server API provides, it must also provide a cancellation method.

- Application servers, which are only needed when certain applications are running, are started when clients need to connect to them. If an application attempts to start a server that is already running, say because it has been started by another application, no error results and only a single instance of the server runs. When the server has no outstanding clients, that is, when the last client session closes, it should terminate to save system resources. This type of server is known as a transient server. I'll illustrate startup and shutdown for this kind of server in the next chapter.
- Other servers, e.g. the POSIX server, are required by only a single application and are started with that application and closed when it terminates.

11.7 How Many Connections Can a Client Have?

A client can have multiple "connections" to a server through one or more sessions as follows:

- Each connection can use a separate client–server session opened by a call to `RSessionBase::CreateSession()`. The sessions are independent of any other within the client thread, and each maintains its own context. That is, each client session has a corresponding `CSharableSession` object in the server and `DSession` object in the kernel. Use of multiple client sessions where they are not strictly necessary should be limited to reduce the number of kernel and server resources consumed – I'll discuss this further later in this chapter.
- The client may create a number of **subsessions** within a single session (the use of subsessions is described in Section 11.14). Client–server communication occurs via the owning session, using a unique handle to identify each individual subsession. The use of separate subsessions is more lightweight than separate sessions because it uses fewer kernel resources. However, they are more complex to implement server-side.
- The server may support sharable sessions. Up to 255 threads in a client process may share a single session.

11.8 What Happens When a Client Disconnects?

Typically, a class used to access a server has a termination method, which is usually called `Close()`. Internally, this method will call `RHandleBase::Close()`, which sends a disconnection message to the server and sets the session handle to zero. On receipt of this message, the server ends its session with the client by destroying the associated `CSharableSession`-derived object (in addition, the kernel will destroy the `DSession` object which represents the session). If the client has any outstanding requests when `Close()` is called, they are not guaranteed to be completed.

11.9 What Happens If a Client Dies?

For a non-sharable session, if the client dies without calling `Close()`, the kernel sends a disconnection message to the server to allow it to cleanup sessions associated with that client thread. The kernel performs its thread-death cleanup by walking the queue of `DSession` objects and destroying any associated with the dead client thread.

If the session is sharable, the death of a single client thread does not close the session – the session is effectively process-relative by virtue of being sharable. To destroy the session, either the client process must terminate or the session must be closed explicitly on the client-side by a call to `Close()` on an `RSessionBase` handle.

11.10 What Happens If a Server Dies?

If a server dies, the kernel will complete any waiting client requests with the error code `KErrServerTerminated`. This gives the client an opportunity to handle request failure and cleanup, destroying any `RSessionBase` objects open on the server. Even if the server is restarted, previous client sessions cannot be reused without first being reconnected to it, so the only valid operation is to call `Close()`.

11.11 How Does Client–Server Communication Use Threads?

A session between a client and a server is between one or more client threads and a separate server thread. Client code runs in user-mode threads. It submits requests to server code which also runs in user mode. The channel of communication between client and server is mediated by the kernel.

The Symbian OS server model is thread-based, allowing a server to run either in a separate process to the client, for greater isolation between client and server, or in the same process, to avoid the overhead of inter-process client–server communication.

11.12 What Are the Implications of Server-Side Active Objects?

The responsiveness of a server can be defined as the maximum time required to process a client message or the maximum time required to respond to an event on some device that it controls. The server uses non-pre-emptive active-object event-handling (described in Chapters 8 and 9). The response time is determined by the longest possible `RunL()` event-handler method of any active object running on the server thread, because an active object cannot be pre-empted when it is handling an event.

If a client makes a request while the server thread is already handling an event in a `RunL()` method, it runs to completion before the client request can be serviced. This is also true for external events occurring from resources owned by the server. Thus, if you want to write a high-performance server, there should be no long-running `RunL()` methods in any active objects in the server's main thread.

This includes processing in the `ServiceL()` method of the `CSharableSession`-derived class, which is called by `CServer::RunL()`. This means that long-running operations must be performed by a separate thread or server.

Furthermore, the priority of a server thread should be chosen according to the maximum guaranteed response time, that is, the longest `RunL()` method of the server. You should not give a high priority to a server thread that performs lots of processing in its event handler, since it may block threads with more appropriately chosen, lower, priorities.

11.13 What Are the Advantages of a Local (Same-Process) Server?

Local servers are useful when several related servers can run in the same process. For example, Symbian OS v7.0 runs the serial communications server, sockets server and telephony server in the same process (`C32.exe`). The servers are in a different process to their clients, so a context switch is still required, and resource integrity is maintained by the separation. However, interactions between the three servers occur in the same process and have a correspondingly lower overhead than they would otherwise (I'll describe the overheads associated with using the client–server model in more detail shortly).

A private local server runs in the same process as its clients. It can be useful, for example, if you need to share client sessions to a server which does not support sharable sessions. The client process should use a private local server which does support sharable sessions and has a single open session with the non-sharable server. This private server services requests from each of the client threads, passing them through as requests to its single session with the non-sharable server.

11.14 What Are the Overheads of Client–Server Communication?

Session Overhead

Although a client can have multiple sessions with a server, each session consumes limited resources in both the server and the kernel. For each open client session, the kernel creates and stores a `DSession` object and the server creates an object of a `CSharableSession`-derived class. This means that each connecting session may give rise to a significant speed overhead. Rather than creating and opening multiple sessions on demand, client code should aim to minimize the number of sessions used. This may involve sharing a session, or, for servers which do not support this, passing the open session between functions or defining classes that store and reuse a single open session.

For efficiency, where multiple sessions are required, a client–server implementation may provide a subsession class to reduce the expense of multiple open sessions. To use a subsession, a client must open a session with the server as normal, and this can then be used to create subsessions which consume fewer resources and can be created more quickly. This is done using the `RSubSessionBase` class, the definition of which is shown below (from `e32std.h`):

```
class RSubSessionBase
    {
public:
    inline TInt SubSessionHandle();
protected:
    inline RSubSessionBase();
    inline RSessionBase& Session();
    IMPORT_C TInt CreateSubSession(RSessionBase& aSession,
            TInt aFunction,const TAny* aPtr);
    IMPORT_C void CloseSubSession(TInt aFunction);
    IMPORT_C TInt Send(TInt aFunction,const TAny* aPtr) const;
    IMPORT_C void SendReceive(TInt aFunction,const TAny* aPtr,
            TRequestStatus& aStatus) const;
    IMPORT_C TInt SendReceive(TInt aFunction,const TAny* aPtr) const;
private:
    RSessionBase iSession;
    TInt iSubSessionHandle;
    };
```

A typical client subsession implementation derives from `RSubSessionBase` in a similar manner to a client session, which derives from `RSessionBase`. The deriving class provides simple wrapper functions to hide the details of the subsession. To open a subsession, the derived class should provide an appropriate wrapper function (e.g. `Open()`) which calls `RSubSessionBase::CreateSubSession()`, passing in an existing `RSessionBase`-derived session object. `CreateSubSession()` also takes an integer "opcode" to identify the "create subsession" request, and a pointer to an array of pointers (which may be used to pass any parameters required to service the request across the client–server boundary).

Once the subsession has been created, `RSubSessionBase::SendReceive()` and `Send()` methods can be called, by analogy with those in `RSessionBase`, but only three parameters of the request data array may be used because the subsession class uses the last element of the data array to identify the subsession to the server.

On the server side, the code to manage client–server subsessions can be quite complex. It usually requires reference counting to manage subsessions over the lifetime of the session, and typically uses the `CObject`-derived classes. You can find more information about these, somewhat confusing, classes in your SDK documentation.

A good example of the use of subsessions is `RFile`, which derives from `RSubSessionBase` and is a subsession of an `RFs` client session to the file server. An `RFile` object represents a subsession for access to individual files. I'll illustrate the use of `RFs` and `RFile` later in this chapter, but you should consult your SDK for further information about the use of the Symbian OS filesystem APIs.

It's worth noting that connections to the file server can take a significant amount of time to set up (up to 75 ms). Rather than creating multiple sessions on demand, `RFs` sessions should be passed between functions where possible, or stored and reused.

Each client–server session has an associated overhead in the kernel and server. Client code should minimize the number of sessions it uses, for example by sharing a session, or by defining classes that store and reuse a single open session. A server may also implement subsessions to be used as lightweight alternatives to multiple open sessions.

Performance Overhead

You should be aware of the system performance implications when using the client–server model. The *amount* of data transferred between the client and server does not cause so much of an overhead as the *frequency* with which the communication occurs.

The main overhead arises because a thread context switch is necessary to pass a message from the client thread to the server thread and back again. If, in addition, the client and server threads are running in different processes, a process context switch is also involved.

A context switch between threads stores the state of the running thread, overwriting it with the previous state of the replacing thread. If the client and server threads are in the same process, the thread context switch stores the processor registers for the threads. If the client and server are running in two separate processes, in addition to the thread context, the process context (the address space accessible to the thread), must be stored and restored. The MMU must remap the memory chunks for each process, and on some hardware this means that the cache must be flushed. The exact nature of the overhead of a thread or process context switch depends on the hardware in question.

Inter-thread data transfer between threads running in separate processes can also have an overhead because an area of data belonging to the client must be mapped into the server's address space.

How Can I Improve Performance?

For performance reasons, when transferring data between the client and server, it is preferable, where possible, to transfer a large amount of data in a single transaction rather than to perform a number of server accesses. However, this must still be balanced against the memory cost associated with storing and managing large blocks of request data.

For example, Symbian OS components that frequently transfer data to or from the filesystem generally do not use direct filesystem access methods such as `RFile::Read()` or `RFile::Write()`. Instead, they tend to use the stream store or relational database APIs, which you can find described in the system documentation. These higher-level components have been optimized to access the file server efficiently. When storing data to file, they buffer it on the client side and pass it to the file server in one block, rather than passing individual chunks of data as it is received.

Thus, taking the stream store for example, `RWriteStream` uses a client-side buffer to hold the data it is passed, and only accesses the file server to write it to file when the buffer is full or if the owner of the stream calls `CommitL()`. Likewise, `RReadStream` pre-fills a buffer from the source file when it is created. When the stream owner wishes to access data from the file, the stream uses this buffer to retrieve the portions of data required, rather than calling the file server to access the file.

When writing code which uses a server, it is always worth considering how to make your server access most efficient. Take the file server, for example: while there are functions to acquire individual directory entries in the filesystem, it is often more efficient to read an entire set of entries and scan them client-side, rather than call across the process boundary to the file server multiple times to iterate through a set of directory entries.

11.15 How Many Outstanding Requests Can a Client Make to a Server?

A client session with a server can only ever have a single synchronous request outstanding. However, it can have up to 255 outstanding asynchronous requests. The message slots which hold these requests are either allocated to a single session or acquired dynamically from a system-wide pool, according to the overload of `RSessionBase::CreateSession()` used to start the session. See the discussion on `RSessionBase` in Section 11.4 for more details.

11.16 Can Server Functionality Be Extended?

Server code can be extended by the use of plug-ins to offer different types of service. A good example of this is the Symbian OS file server, which can be extended at runtime to support different types of filesystem plug-in. The core Symbian OS filesystem provides support for local media (ROM, RAM and CF card) using a VFAT filing system, implemented as a plug-in, `ELocal.fsy`. Other filesystems may be added, for example to support a remote filesystem over a network or encryption of file data before it is stored. These file system plug-ins may be installed and uninstalled dynamically; in the client-side file server session class, `RFs`, you'll see a set of functions for this purpose (`FileSystemName()`, `AddFileSystem()`, `RemoveFileSystem()`, `MountFileSystem()` and `DismountFileSystem()`).

The extension code should be implemented as a polymorphic DLL (`targettype fsy`) and must conform to the `fsy` plug-in interface defined by Symbian OS. More information about the use of framework and plug-in code in Symbian OS, and polymorphic DLLs, can be found in Chapter 13.

Since they normally run in the same thread as the server, and are often called directly from `CServer::RunL()`, it is important to note that installable server extension plug-in modules must not have a negative impact on the performance or runtime stability of the server.

11.17 Example Code

Finally, here is an example of how a server may be accessed and used. It illustrates how a client thread creates a session with the Symbian OS file server. The file server session class, `RFs`, is defined in `f32file.h`, and to use the file server client-side implementation you must link against `efsrv.lib`.

Having successfully created the session and made it leave-safe using the cleanup stack (as described in Chapter 3), the sample code submits a request to the file server, using `RFs::Delete()`. This function wraps the single descriptor parameter and passes it to the synchronous overload of `RSessionBase::SendReceive()`. Following this, it creates an `RFile` subsession of the `RFs` session, by calling `RFile::Create()`. It then calls `RFile::Read()` on the subsession, which submits a request to the file server by wrapping a call to `RSubSessionBase::SendReceive()`, passing in the descriptor parameter that identifies the buffer which will receive data read from the file.

```
RFs fs;
User::LeaveIfError(fs.Connect());
CleanupClosePushL(fs); // Closes the session in the event of a leave

_LIT(KClangerIni, "c:\\clanger.ini");

// Submits a delete request to the server
User::LeaveIfError(fs.Delete(KClangerIni));

RFile file; // A subsession
User::LeaveIfError(file.Create(fs, KClangerIni,
        EFileRead|EFileWrite|EFileShareExclusive));
// Closes the subsession in the event of a leave
CleanupClosePushL(file);

TBuf8<32> buf;
// Submits a request using the subsession
User::LeaveIfError(file.Read(buf));
CleanupStack::PopAndDestroy(2, &fs); // file, fs
```

11.18 Summary

This chapter explored the theory behind the Symbian OS client–server framework. It is quite a complex subject, so the chapter was split into a number of short sections to allow it to be read at several different levels, depending on whether the information is required to *use* a server or to *write* one. The sections covered the following:

- The basics of the client–server framework and why it is used to share access to system resources and protect their integrity.
- The thread model for a client–server implementation. The client and server run in separate threads and often the server runs in a separate process; this isolates the system resource within a separate address space and protects it from potential misuse. Symbian OS threads and processes are discussed in Chapter 10.

- The mechanism by which data is transferred between the client and server, using inter-thread data transfer. This uses inter-process communication (IPC) when the client and server run in different processes, and there are potential runtime overheads associated with IPC data transfer.
- The main classes which make up the Symbian OS client–server framework (`RSessionBase`, `C(Sharable)Session`, `CServer`, `DSession`, `RMessage` and `RSubSessionBase`), their main roles, features, fundamental methods and interactions.
- Server startup (which is examined in more detail in the next chapter) and shutdown, either when all clients have disconnected normally or as a result of the death of either the client or server.
- The overheads associated with using a client–server architecture, such as the number of kernel resources consumed by having multiple open sessions with a server, and the potential impact of using active objects with lengthy `RunL()` event handlers within the server.
- Example code using F32, the file server, as an example of a typical Symbian OS client–server model.

The following chapter uses a detailed code example to illustrate a typical server, its client-side implementation, and its use by a calling client.

12

The Client–Server Framework in Practice

Kill the lion of Nemea
Kill the nine-headed Hydra
Capture the Ceryneian Hind
Kill the wild boar of Erymanthus
Clean the stables of King Augeas
Kill the birds of Stymphalis
Capture the wild bull of Crete
Capture the man-eating mares of Diomedes
Obtain the girdle of Hippolyta, Queen of the Amazons
Capture the oxen of Geryon
Take golden apples from the garden of Hesperides
Bring Cerberus, the three-headed dog of the underworld, to the surface
The Twelve Labors of Hercules

This chapter works through the code for an example client and server to illustrate the main points of the Symbian OS client–server architecture, which I discussed in detail in Chapter 11. This chapter will be of particular interest if you plan to implement your own server, or if you want to know more about how a client's request to a server is transferred and handled.

The code examines the typical features of a client–server implementation using a transient server, which is started by its first client connection and terminates when its last outstanding client session closes, to save system resources. I'll take the main elements of client–server code in turn and discuss the most important sections of each. You can find the entire set of sample code on the Symbian Press website (***www.symbian.com/books***). The bootstrap code used by a client to start a server can be quite complex and you may find it helps to download this example and step through the code to follow how it works.

As in the previous chapter, I discuss the client–server model only for Symbian OS releases up to and including v7.0s (the code samples in this chapter use the client–server APIs from Symbian OS v7.0). Some of the APIs have changed from Symbian OS 8.0 onwards and, although the concepts are generally the same, I've decided to concentrate solely on the releases available at the time of going to press, rather than confuse matters.

Client–server code can be notoriously complicated, so I have kept the services provided by the server as simple as possible. This comes, perhaps, at the expense of making the example code rather contrived; my client–server example provides a software representation of the twelve labors of the Greek hero Hercules.[1] For each labor, I've exported a function from a client-side implementation class, `RHerculesSession`. These functions send a request to the server to perform the necessary heroic activity. The client-side implementation is delivered as a shared library DLL (`client.dll`) – callers wishing to use the functionality provided by server should link to this. The server code itself is built into a separate `EPOCEXE` component (shared library DLLs and `targettype EPOCEXE` are discussed in Chapter 13).

12.1 Client–Server Request Codes

A set of enumerated values is used to identify which service the client requests from the server. These values are quite straightforward and are defined as follows:

```
enum THerculeanLabors
    {
    ESlayNemeanLion,
    ESlayHydra,
    ECaptureCeryneianHind,
    ESlayErymanthianBoar,
    ECleanAugeanStables,
    ESlayStymphalianBirds,
    ECaptureCretanBull,
    ECaptureMaresOfDiomedes,
    EObtainGirdleOfHippolyta,
    ECaptureOxenOfGeryon,
    ETakeGoldenApplesOfHesperides,
    ECaptureCerberus,
    ECancelCleanAugeanStables,
    ECancelSlayStymphalianBirds
    };
```

Later in the chapter you'll see these shared request "opcodes" passed to an overload of `RSessionBase::SendReceive()` on the client side and stored in the corresponding server-side `RMessage` object (which represents the client request) to identify which service the client has requested.

[1] If nothing else, this chapter will prepare you well for a pub quiz question about the Herculean labors.

12.2 Client Boilerplate Code

Much of the client-side implementation is made up of the API used by callers to submit requests to the server. This API is exported by `RHerculesSession`, which derives from `RSessionBase`. Each of the request methods passes the associated opcode, and any parameter data, to the server via a call to the base class method `RSessionBase::SendReceive()`, using the synchronous or asynchronous overload as appropriate.

Here is the definition of the main client-side class (I've shown only six of the Herculean labor request methods):

```
// Forward declarations - the actual class declarations must be
// accessible to both client and server code
class CHerculesData;
struct THydraData;

class RHerculesSession : public RSessionBase
    {
public:
    IMPORT_C TInt Connect();
public:
    IMPORT_C TInt SlayNemeanLion(const TDesC8& aDes, TInt aVal);
    IMPORT_C TInt SlayHydra(THydraData& aData);
    IMPORT_C TInt CaptureCeryneianHind(TInt& aCaptureCount);
    IMPORT_C TInt SlayErymanthianBoar(const CHerculesData& aData);
    IMPORT_C void CleanAugeanStables(TRequestStatus& aStatus);
    IMPORT_C void CancelCleanAugeanStables();
    IMPORT_C void SlayStymphalianBirds(TInt aCount, TDes8& aData,
            TRequestStatus& aStatus);
    IMPORT_C void CancelSlayStymphalianBirds();
    ...
    };
```

I've included a range of parameter input, and implemented synchronous and asynchronous functions for illustration purposes. Here are the implementations of the request submission methods:

```
EXPORT_C TInt RHerculesSession::SlayNemeanLion(const TDesC8& aDes,
        TInt aVal)
    {
    const TAny* p[KMaxMessageArguments];
    p[0]=&aDes;
    p[1]=(TAny*)aVal;
    return (SendReceive(ESlayNemeanLion,p));
    }

EXPORT_C TInt RHerculesSession::SlayHydra(THydraData& aData)
    {
    const TAny* p[KMaxMessageArguments];
    TPckg<THydraData> data(aData);
    p[0]=&data;
    return (SendReceive(ESlayHydra,p));
```

```
    }

EXPORT_C TInt RHerculesSession::CaptureCeryneianHind(TInt& aCaptureCount)
    {
    const TAny* p[KMaxMessageArguments];
    TPckg<TInt> countBuf(aCaptureCount);
    p[0]=(TAny*)&countBuf;
    return (SendReceive(ECaptureCeryneianHind,p));
    }

// The implementation of RHerculesSession::SlayErymanthianBoar()
// is omitted here because it is discussed later in the section

// Asynchronous request
EXPORT_C void RHerculesSession::CleanAugeanStables(TRequestStatus& aStat)
    {
    SendReceive(ECleanAugeanStables, 0, aStat);
    }
// Cancels the CleanAugeanStables() asynchronous request
EXPORT_C void RHerculesSession::CancelCleanAugeanStables()
    {
    SendReceive(ECancelCleanAugeanStables, 0);
    }

// Asynchronous request
EXPORT_C void RHerculesSession::SlayStymphalianBirds(TInt aCount,
        TDes8& aData, TRequestStatus& aStatus)
    {
    const TAny* p[KMaxMessageArguments];
    p[0] = (TAny*)aCount;
    p[1] = &aData;
    SendReceive(ESlayStymphalianBirds, p, aStatus);
    }
// Cancels the SlayStymphalianBirds() asynchronous request
EXPORT_C void RHerculesSession::CancelSlayStymphalianBirds ()
    {// Every asynchronous request should have a cancellation method
    SendReceive(ECancelSlayStymphalianBirds, 0);
    }
```

You'll understand now why I've called it "boilerplate" code – there's a lot of repetition in the methods.

In each case, you'll notice that, if any parameter data is passed to the server with the request, the method instantiates an array of 32-bit values of size `KMaxMessageArguments` (= 4, as defined in `e32std.h`). If there are no accompanying request parameters (as in the case of the request with opcode `ECleanAugeanStables` or the cancellation methods), the array is not needed. The array is used to hold either the request data itself (if it can be stored in the 32-bit elements) or a pointer to a descriptor that stores the client-side data. It is the contents of this array that are stored in an `RMessage` on the server side, but there is no direct equivalent of `RMessage` for client-side code. The array is passed to `RSessionBase::SendReceive()` and must always be of size `KMaxMessageArguments` even when fewer parameters are passed to the server.

It is important that the client-side data passed to an asynchronous request must not be stack-based. This is because the server may not process the incoming request data until some arbitrary time after the client issued the request. The parameters must remain in existence until that time – so they cannot exist on the stack in case the client-side function which submitted the request returns, destroying the stack frame.

The client API differentiates between non-modifiable descriptors passed to functions which pass constant data to the server and modifiable descriptors used to retrieve data from it. However, the client-side interface code simply passes a pointer to the descriptor as a message parameter. Server-side, this will be used in a call to `RThread::ReadL()` to retrieve data from the client or in a call to `RThread::WriteL()` to write data to the client thread. The `RThread` methods inspect the descriptor to which it points, to check that it appears to be a descriptor, and leave with `KErrBadDescriptor` if it does not.

`SlayNemeanLion()` and `CaptureCeryneianHind()` show how integer and descriptor data are passed to a server, but what about custom data? What if it has variable length or does not just contain "flat" data, but owns pointers to other objects, as is common for a C class object? I'll show how to pass a `CBase`-derived object across the client–server boundary in `SlayErymanthianBoar()` shortly, but first, let's consider how to pass an object of a T class or a `struct`.

`RHerculesSession::SlayHydra()` passes an object of type `THydraData` which is a simple `struct` that contains only built-in types, defined as follows:

```
struct THydraData
    {
    TVersion iHydraVersion;
    TInt iHeadCount;
    };
```

`TVersion` is a Symbian OS class defined as follows in `e32std.h`:

```
class TVersion
    {
public:
    ... // Constructors omitted for clarity
public:
    TInt8 iMajor;
    TInt8 iMinor;
    TInt16 iBuild;
    };
```

A `THydraData` object is thus 64 bits in size, which is too large to be passed to the server as one of the 32-bit elements of the request data array. It isn't enough to pass a pointer to the existing `THydraData` object either because, when the client and server are running in different processes, the server code runs in a different virtual address space. Under these circumstances, a C++ pointer which is valid in the client process is not valid in the server process; any attempt to use it server-side will result in an access violation.[2]

Server-side code should not attempt to access a client-side object directly through a C++ pointer passed from the client to server. Data transfer between client and server *must* be performed using the inter-thread data transfer[3] methods of class `RThread` – except when integer or boolean values are passed from the client. The request data array can be used to pass up to four 32-bit values to the server, so these can be read directly from the request message server-side. However, if the parameter is a reference which the server updates for the client (such as in the `CaptureCeryneianHind()` method above), the server must use a kernel-mediated transfer to write the 32-bit data back to the client. I'll use example code to illustrate this later in the chapter.

`RHerculesSession::SlayNemeanLion()` transfers a descriptor parameter (`const TDesC8&aDes`) from the client to the server by passing a pointer to that descriptor as one of the elements of the request data array. However, `THydraData` is not a descriptor and before it is passed to the server it must be "descriptorized". Chapter 6 discusses the use of the package pointer template classes `TPckg` and `TPckgC`, which can be used to wrap a flat data object such as `THydraData` with a pointer descriptor, `TPtr8`. The `SlayHydra()` method of `RHerculesSession` creates a `TPckg<THydraData>` around its `THydraData` parameter to pass to the server in the request data array. The resulting descriptor has a length equivalent to the size in bytes of the templated object it wraps, and the `iPtr` data pointer of the `TPtr8` addresses the start of the `THydraData` object. Later in the chapter I'll show how the server accesses the data and retrieves the `THydraData` object.

Moving on, let's consider how an object of a C class, containing a pointer to another object or variable-length data, is marshaled from client to server. Consider the `CHerculesData` class, which owns two heap descriptor pointers and an integer value. We've already seen that passing a client-side object to the server requires the entire object to be "descriptorized" for passing across the process boundary. However, as I've already described, pointers to data in the client address space cannot

---

[2] It should be noted that standard C++ pointer access directly between the client and server may succeed on the Windows emulator. The emulator runs as a single process with each Symbian OS process running as a separate thread. Threads share writable memory and therefore the address spaces of the client and server are not separated by a process boundary, but are mutually accessible. However, this will most definitely not be the case when Symbian OS runs on real phone hardware, so you must not use direct pointer access to transfer data between the client and server processes.

[3] Where the client and server are running in different processes this transfer is, in effect, inter-process communication (IPC).

be used server-side. Thus, for the server to use a `CHerculesData` object, it must receive a copy of the data each heap descriptor pointer addresses.

The `CHerculesData` class must have utility code which puts all its member data into a descriptor client-side ("externalization") and corresponding code to recreate it from the descriptor server-side ("internalization"). There is a standard technique for this, as shown below:

```
class CHerculesData : public CBase
    {
public:
    IMPORT_C static CHerculesData* NewLC(const TDesC8& aDes1,
            const TDesC8& aDes2, TInt aVal);
    static CHerculesData* NewLC(const TDesC8& aStreamData);
    IMPORT_C ~CHerculesData();
    ... // Other methods omitted
public:
    // Creates an HBufC8 representation of 'this'
    IMPORT_C HBufC8* MarshalDataL() const;
protected:
    // Writes 'this' to the stream
    void ExternalizeL(RWriteStream& aStream) const;
    // Initializes 'this' from stream
    void InternalizeL(RReadStream& aStream);
protected:
    CHerculesData(TInt aVal);
    CHerculesData(){};
    void ConstructL(const TDesC8& aDes1, const TDesC8& aDes2);
protected:
    HBufC8* iDes1;
    HBufC8* iDes2;
    TInt iVal;
    };

// Maximum size expected for iDes1 and iDes2 in CHerculesData
const TInt KMaxHerculesDesLen = 255;

// Maximum total size expected for a CHerculesData object
const TInt KMaxCHerculesDataLength = 520;

EXPORT_C CHerculesData* CHerculesData::NewLC(const TDesC8& aDes1,
            const TDesC8& aDes2, TInt aVal)
    {
    CHerculesData* data = new (ELeave) CHerculesData(aVal);
    CleanupStack::PushL(data);
    data->ConstructL(aDes1, aDes2);
    return (data);
    }

// Creates a CHerculesData initialized with the contents of the
// descriptor parameter
CHerculesData* CHerculesData::NewLC(const TDesC8& aStreamData)
    {// Reads descriptor data from a stream
     // and creates a new CHerculesData object
    CHerculesData* data = new (ELeave) CHerculesData();
    CleanupStack::PushL(data);
    // Open a read stream for the descriptor
```

```
    RDesReadStream stream(aStreamData);
    CleanupClosePushL(stream);
    data->InternalizeL(stream);
    CleanupStack::PopAndDestroy(&stream); // finished with the stream
    return (data);
    }

EXPORT_C CHerculesData::~CHerculesData()
    {
    delete iDes1;
    delete iDes2;
    }

CHerculesData::CHerculesData(TInt aVal)
    :iVal(aVal){}

void CHerculesData::ConstructL(const TDesC8& aDes1, const TDesC8& aDes2)
    {
    iDes1 = aDes1.AllocL();
    iDes2 = aDes2.AllocL();
    }

// Creates and returns a heap descriptor which holds contents of 'this'
EXPORT_C HBufC8* CHerculesData::MarshalDataL() const
    {
    // Dynamic data buffer
    CBufFlat* buf = CBufFlat::NewL(KMaxCHerculesDataLength);
    CleanupStack::PushL(buf);
    RBufWriteStream stream(*buf); // Stream over the buffer
    CleanupClosePushL(stream);
    ExternalizeL(stream);
    CleanupStack::PopAndDestroy(&stream);

    // Create a heap descriptor from the buffer
    HBufC8* des = HBufC8::NewL(buf->Size());
    TPtr8 ptr(des->Des());
    buf->Read(0, ptr, buf->Size());
    CleanupStack::PopAndDestroy(buf); // Finished with the buffer
    return (des);
    }

// Writes 'this' to aStream
void CHerculesData::ExternalizeL(RWriteStream& aStream) const
    {
    if (iDes1) // Write iDes1 to the stream (or a NULL descriptor)
        aStream << *iDes1;
    else
        aStream << KNullDesC8;

    if (iDes2) // Write iDes2 to the stream (or a NULL descriptor)
        aStream << *iDes2;
    else
        aStream << KNullDesC8;

    aStream.WriteInt32L(iVal); // Write iVal to the stream
    }

// Initializes 'this' with the contents of aStream
```

```
void CHerculesData::InternalizeL(RReadStream& aStream)
    {
    iDes1 = HBufC8::NewL(aStream, KMaxHerculesDesLength); // Read iDes1
    iDes2 = HBufC8::NewL(aStream, KMaxHerculesDesLength); // Read iDes2
    iVal = aStream.ReadInt32L();                           // Read iVal
    }
```

`CHerculesData` has a public method `MarshalDataL()`, which creates and returns a heap descriptor containing the current contents of the object. The method creates a dynamic buffer and passes a writable stream over this buffer to its `ExternalizeL()` method. The protected externalization method then writes the data of each individual member of `CHerculesData` to the stream, using the built-in stream externalization support provided by Symbian OS. `ExternalizeL()` uses the stream `operator<<` to write the heap descriptors to the stream (as I discussed in Chapter 6).

Having created an in-memory stream of the contents of a `CHercules-Data` object in the dynamic buffer, the `MarshalDataL()` method then converts it to a heap descriptor by allocating a descriptor of the appropriate size and copying the data from the dynamic buffer.

Server-side, the descriptor data is read from the client thread by making an inter-thread data transfer. To convert the descriptorized `CHercules-Data` back to an object of that class, the server calls the overload of `CHerculesData::NewLC()` passing the descriptor which contains the descriptorized object. This method creates a descriptor read stream and calls `InternalizeL()`, which is the opposite of `ExternalizeL()`, and instantiates the heap descriptor member variables from the stream.

Any class that holds variable-length data, such as an `RArray`-derived object or pointers to heap-based objects, must provide similar externalize and internalize functions if objects of that type need to be marshaled between a client and server.

Here's the code for the `RHerculesSession` method which marshals a `CHerculesData` object into a descriptor and passes it from client to server using `RSessionBase::SendReceive()`:

```
EXPORT_C TInt RHerculesSession::SlayErymanthianBoar(const CHerculesData&
        aData)
    {
    const TAny* p[KMaxMessageArguments];
    HBufC8* dataDes;
    TRAPD(r, dataDes = aData.MarshalDataL());
    if (dataDes)
        {
        p[0] = dataDes;
        r = SendReceive(ESlayErymanthianBoar, p);
        delete dataDes;
        }
    return (r);
    }
```

Figure 12.1 illustrates the previous discussion and code sample.

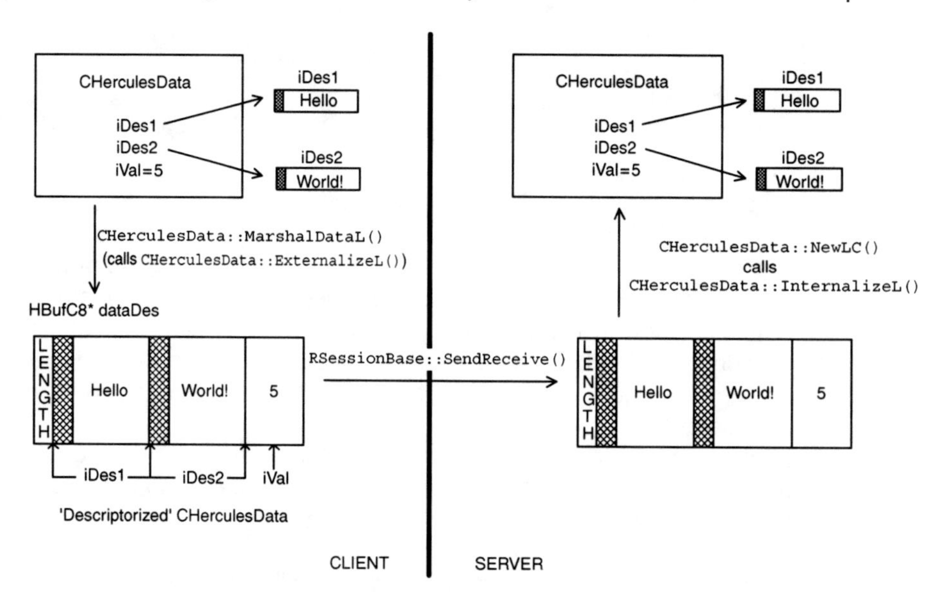

Figure 12.1 Marshaling `CHerculesData` from client to server within a descriptor

A client and server typically run in different processes. For this reason, server-side code should not attempt to access a client-side object directly through a C++ pointer passed from the client to server. Where necessary, objects must be "descriptorized" for inter-process communication.

12.3 Starting the Server and Connecting to It from the Client

The `RHerculesSession` constructor zeroes its `iHandle` value so it is clear that the `RHerculesSession` object is invalid and must first connect to the server by calling `RSessionBase::CreateSession()`. The client-side implementation of server access code typically wraps this call in a method called `Connect()`, e.g. `RFs::Connect()`, or occasionally an `Open()` method. If an attempt is made to submit a request to a server using a client session that has not yet connected to the server, a panic occurs (`KERN-EXEC 0`).

If the server is an essential system server, it is guaranteed to have been started by the system as Symbian OS starts and will always be running. The definition of a system server, for example, the file server, is that it is required by the system. If a system server panics for some reason, Symbian OS is forced to reboot and restart the server because it cannot function without it.

Client connection to a system server is straightforward because the server is already running. In effect, a Connect() method simply needs to call RSessionBase::CreateSession(). However, as I've described, the Hercules example server discussed in this chapter is not an essential system server; instead it is implemented as a transient server. If it is not already running when a client calls CreateSession(), that is if no other client is connected to it, the Connect() method of the client-side implementation must start the server process. The server runs in a separate process on phone hardware and Symbian OS processes run in separate threads within the single process of the Windows emulator.

Here is the implementation of Connect() for the Hercules server:

```
// The server's identity within the client - server framework
_LIT(KServerName,"HerculesServer");

EXPORT_C TInt RHerculesSession::Connect()
    {
    TInt retry=2; // A maximum of two iterations of the loop are required
    for (;;)
        {
        // Uses system-pool message slots
        TInt r=CreateSession(KServerName,TVersion(1,0,0));
        if ( (KErrNotFound!=r) && (KErrServerTerminated!=r) )
            return (r);

        if (--retry==0)
            return (r);

        r=StartTheServer();
        if ( (KErrNone!=r) && (KErrAlreadyExists!=r) )
            return (r);
        }
    }
```

The method calls RSessionBase::CreateSession(), passing in the name of the server required. If the server is already running and can create a new client session, this call returns KErrNone and the function returns successfully, having connected a new session. However, if the server is not running, CreateSession() returns KErrNotFound. The Connect() method must check for this result and, under these circumstances, attempt to start the server. You'll notice that the method also checks to see if KErrServerTerminated was returned instead of KErrNotFound, which indicates that the client connection request was submitted just as the server was shutting down. For either value,[4] the server should be started (or restarted) by calling

[4] If any other error value besides KErrNotFound or KErrTerminated is returned, Connect() returns it to the caller.

StartTheServer(). I'll discuss this function shortly – if it returns KErrNone, the server has started successfully. The next iteration of the for loop submits another request to create a client session by calling RSessionBase::CreateSession().

If some other client managed to start the server between this client's failed attempt to create a session on it and the call to StartTheServer(), it returns KErrAlreadyExists. Under the circumstances, this is not an error,[5] because the server is now running, even if the client didn't start it. So, again, the next iteration of the loop can submit a request to create a new client session.

If the server fails to start or the client connection fails, an error is returned and the code breaks out of the loop, returning the error value to indicate that session creation failed.

StartTheServer() launches the server in a new thread on the Windows emulator, or in a new, separate process on hardware running Symbian OS. Because the server runs differently depending on whether it is running on the emulator or a phone, the client-side server startup code is quite complex. The code is implemented as follows (I've included a number of comments to make it clear what is going on and I'll discuss the most important features below):

```
// Runs client-side and starts the separate server process/thread
static TInt StartTheServer()
    {
    TRequestStatus start;
    TServerStart starter(start);

    // HerculesServer.exe or HerculesServer.dll
    _LIT(KServerBinaryName,"HerculesServer");
    const TUid KServerUid3={0x01010101}; // Temporary UID
    const TUidType serverUid(KNullUid, KNullUid, KServerUid3);

#ifdef __WINS__ // On the Windows emulator the server is a DLL
    RLibrary lib;
    TInt r=lib.Load(KServerBinaryName, serverUid);
    if (r!=KErrNone)
        return (r);
    // The entry point returns a TInt representing the thread
    // function for the server
    TLibraryFunction export1 = lib.Lookup(1);
    TThreadFunction threadFunction =
                reinterpret_cast<TThreadFunction>(export1());
    TName name(KServerName); // Defined previously
    // Randomness ensures a unique thread name
    name.AppendNum(Math::Random(), EHex);
```

[5] The KErrAlreadyExists error code is returned because there must only be one copy of the server running on the system. Simultaneous attempts to launch two copies of the server process will be detected by the kernel when the second attempt is made to create a CServer object of the same name. It is this which fails with KErrAlreadyExists.

```
    // Now create the server thread
    const TInt KMinServerHeapSize=0x1000;
    const TInt KMaxServerHeapSize=0x1000000;

    RThread server;
    r = server.Create(name, threadFunction,
                KDefaultStackSize, &starter, &lib, NULL,
                KMinServerHeapSize, KMaxServerHeapSize, EOwnerProcess);

    lib.Close(); // The server thread has a handle to the library now

#else                   // Phone hardware - comparatively easy -
    RProcess server;    // just create a new process
    TInt r=server.Create(KServerBinaryName,starter.AsCommand(),
            serverUid);
#endif

    if (KErrNone!=r)
        return r;

    TRequestStatus threadDied;
    server.Logon(threadDied);
    if (KRequestPending!=threadDied)
        {// logon failed - server isn't running yet
        // Manage the thread signal by consuming it
        User::WaitForRequest(threadDied);
        server.Kill(0); // don't try to start the server
        server.Close();
        return threadDied.Int(); // return the error
        }

    // Start the server thread or process and wait for startup or thread
    // death notification
    // This code is identical for both the emulator thread
    // and the phone hardware process
    server.Resume();
    User::WaitForRequest(start, threadDied);
    if (KRequestPending==start)
        {// server died and signaled - startup still pending
        server.Close();
        return threadDied.Int();
        }
// The server started.
// Cancel the logon and consume the cancellation signal
    server.LogonCancel(threadDied);
    server.Close(); // Don't need this handle any more
    User::WaitForRequest(threadDied);
    return (KErrNone);
    }
```

As you can see, `StartTheServer()` is responsible for creating the new thread or process in which the server runs. Having done so, it waits for notification that the server thread or process has been initialized and returns the initialization result to its caller. You'll notice that the client waits for server notification by calling `User::WaitForRequest()`, which means that the client thread is blocked until the server responds. For this reason, server initialization should be as rapid as possible, performing only the actions necessary to initialize the server framework. Other initialization should be performed asynchronously within the server after it has been successfully launched.

On the emulator, the new thread is created with a unique name, to avoid receiving a `KErrAlreadyExists` error from the kernel if the server is restarted just after it has exited. The thread is assigned a standard stack size, minimum and maximum heap sizes and an `RLibrary` handle to the DLL in which the server code resides. The thread entry function is initialized to the first export function of the DLL, which will be discussed shortly when I come to describe the server-side startup code. The server thread is owned by the entire emulator process (as indicated by the `EOwnerProcess` parameter) rather than by the client thread that creates it. This means that the client thread can terminate but the server thread will continue to run, allowing other client threads to access the server without it dying unexpectedly.

On a real Symbian OS phone, the server is launched as a process, the code for which is significantly more straightforward than for emulator thread startup. Once the server thread or process has started, the client thread calls `Logon()` in case a failure occurs before server initialization function is complete. The `Logon()` call ensures that the client thread is notified if the server dies before the server startup `TRequestStatus` is signaled. This prevents the client thread from hanging indefinitely if server initialization fails. Once the server entry point signals the client and sets the value of `start` to something other than `KRequestPending`, the `Logon()` notification is canceled. I'll discuss how the server-side code manages initialization later in this chapter.

The client thread passes a `TRequestStatus` to the server for notification when its initialization is complete. A `TServerStart` object ("`starter`") is used to transfer the client's `TRequestStatus` object ("`start`") to the server. The `TServerStart` class is defined and implemented as follows:

```
class TServerStart
    {
public:
    TServerStart() {};
    TServerStart(TRequestStatus& aStatus);
    TPtrC AsCommand() const;
    TInt GetCommand();
    void SignalL();
private:
    TThreadId iId;
    TRequestStatus* iStatus;
    };

inline TServerStart::TServerStart(TRequestStatus& aStatus)
:iId(RThread().Id()),iStatus(&aStatus){aStatus=KRequestPending;}
```

```
// Descriptorizes 'this' to pass it from client to server
inline TPtrC TServerStart::AsCommand() const
    {return TPtrC(reinterpret_cast<const TText*>(this),
            sizeof(TServerStart)/sizeof(TText));}

// Called server-side to 'reconstitute' an object of this type
TInt TServerStart::GetCommand()
    {
    RProcess serverProcess;
    if (serverProcess.CommandLineLength()! =
            sizeof(TServerStart)/sizeof(TText))
        return (KErrGeneral);

    TPtr ptr(reinterpret_cast<TText*>(this),0,
            sizeof(TServerStart)/sizeof(TText));
    serverProcess.CommandLine(ptr);
    return (KErrNone);
    }

// Called server-side to notify the client that server initialization
// completed successfully
void TServerStart::SignalL()
    {
    RThread client;
    User::LeaveIfError(client.Open(iId));
    client.RequestComplete(iStatus, KErrNone);
    client.Close();
    }
```

A `TServerStart` object encapsulates the client's `TThreadId` and a pointer to a `TRequestStatus` object. For hardware builds, the object is passed between the client and server processes by wrapping it in a descriptor, using the `AsCommand()` method and passing it as a command line argument. Server-side, a `TServerStart` object can retrieve these values using `GetCommand()`, which reads the command line with which the process was created. On the emulator, the `TServerStart` object is passed as a parameter to the thread function. The server uses `TServerStart::SignalL()` to notify the client that initialization has completed successfully.

The code to start a server is quite complex because it differs depending on whether the server will run on a phone or the emulator. This complexity has been removed from Symbian OS v8.0 because Symbian OS process emulation on Win32 has been improved (see Chapter 10 for more details).

12.4 Server Startup Code

Having considered the client-side code to start the server, I will now move on to discuss the server-side startup code:

```
// Initialize and run the server
static void RunTheServerL(TServerStart& aStart)
    {// First create and install the active scheduler
    CActiveScheduler* scheduler = new (ELeave) CActiveScheduler;
    CleanupStack::PushL(scheduler);
    CActiveScheduler::Install(scheduler);

    CHerculeanServer::NewLC();// creates the server
    User::LeaveIfError(RThread().Rename(KServerName));

    aStart.SignalL();// Signal the client that initialization is complete

    CActiveScheduler::Start();// Enter the wait loop
    // Exited - cleanup the server and scheduler
    CleanupStack::PopAndDestroy(2, scheduler);
    }

// Main entry-point for the server thread/process
static TInt RunTheServer(TServerStart& aStart)
    {
    CTrapCleanup* cleanup=CTrapCleanup::New();
    TInt r=KErrNoMemory;
    if (cleanup)
        {
        TRAP(r,RunTheServerL(aStart));
        delete cleanup;
        }
        return (r);
    }

#ifdef __WINS__ // Different startup code for emulator and hardware

// Thread entry-point function
static TInt ThreadFunction(TAny* aParameters)
    {// The TServerStart object is passed as the thread parameter
    return RunTheServer(*static_cast<TServerStart*>(aParameters));
    }

// WINS DLL entry-point
IMPORT_C TInt WinsMain();
EXPORT_C TInt WinsMain()
    {// Returns the real thread function cast to TInt
    return reinterpret_cast<TInt>(&ThreadFunction);
    }

TInt E32Dll(TDllReason)
    {return (KErrNone);}

#else // Phone hardware
// Process entry-point
TInt E32Main()
    {// Reads the startup parameters and runs the server
    TServerStart start;
    TInt r=start.GetCommand();
    if (KErrNone==r)
    r=RunTheServer(start);
    return (r);
    }
#endif
```

Again, as you can see, the startup code differs for hardware process and emulator DLL builds. On the emulator, the server DLL has an entry point called `WinsMain()`, which takes no parameters and returns a `TInt`. However, the Symbian OS thread function entry point function must take a `TAny*` parameter, so `WinsMain()` is used to return the actual thread function (`ThreadFunction()`) by casting the function pointer to a `TInt`. The `TAny*` parameter passed to the thread function refers to the client's `TServerStart` object. This is passed directly to the main server startup function, called `RunTheServer()`. It sounds quite complex, but as you can see, it only takes a few lines of code.

On hardware, the server is a process with `E32Main()` as its entry point. It simply instantiates a `TServerStart` object and uses `GetCommand()` to retrieve the client's `TThreadId` and `TRequestStatus`. Having done so, it calls `RunTheServer()`. At this point, the divergence between the emulator and hardware code is at an end.

`RunTheServer()` first creates a cleanup stack for the server. Having done so, it calls `RunTheServerL()` within a `TRAP` harness (recall from Chapter 2 that there must be at least one top-level `TRAP` in the system before the cleanup stack can be used). `RunTheServerL()` creates and installs an active scheduler to allow it to service incoming client requests. It then creates the `CServer`-derived object (and any objects that uses) and calls `CServer::Start()` or `CServer::StartL()`. Once the server object is created and ready to receive connection requests, the server thread is renamed with the name of the server. This is an optional step, but it is useful when debugging panics to be able to see the name of the server thread. At this point, server startup is complete and `TServerStart::SignalL()` is called to notify the client that the server is fully initialized.

Having signaled the client, the `Start()` method is called on the active scheduler to make it ready to receive and process client requests. Within this function, the server thread enters the active scheduler wait loop, receiving and servicing client requests until `CActive-Scheduler::Stop()` is called.

12.5 Server Classes

The fundamental server-side classes, deriving from `CServer` and `CSharableSession`, are defined as follows:

```
class CHerculeanServer : public CServer // Receives client requests
    {
public:
    static CServer* NewLC();
    void AddSession();
    void RemoveSession();
```

```
protected:
    virtual TInt RunError(TInt aError); // Overrides CActive::RunError()
private:
    CHerculeanServer();
    void ConstructL();
    virtual CSharableSession* NewSessionL(const TVersion& aVersion)
            const; // From CServer
private:
    TInt iSessionCount;
    CShutdown iShutdown;
    };

inline CHerculeanServer::CHerculeanServer()
:CServer(0, ESharableSessions){}

class CAsyncHandler; // Active object class for asynchronous requests

// Services client requests
class CHerculeanSession : public CSharableSession
    {
public:
    CHerculeanSession(){};
    virtual void CreateL(const CServer& aServer);
private:
    void SlayNemeanLionL(const RMessage& aMessage);
    void SlayHydraL(const RMessage& aMessage);
    void CaptureCeryneianHindL(const RMessage& aMessage);
    void SlayErymanthianBoarL(const RMessage& aMessage);
    void CleanAugeanStablesL(const RMessage& aMessage);
    void SlayStymphalianBirdsL(const RMessage& aMessage);
private:
    ~CHerculeanSession();
    inline CHerculeanServer& Server();
    void ServiceL(const RMessage& aMessage); // From CSharableSession
private:
    CAsyncHandler* iAsyncRequestHandler;
    HBufC8* iClientBuf;
    };

inline CHerculeanServer& CHerculeanSession::Server()
    {
    return *static_cast<CHerculeanServer*>(const_cast<CServer*>
            (CSharableSession::Server()));
    }
```

The `CServer`-derived class `CHerculeanServer` is the main server class; it coordinates server startup and shutdown and, as I described in the previous chapter, receives incoming requests. The system creates a single instance of `CHerculeanServer` on server startup. The `CServer` base class manages a doubly-linked list of connected client sessions. When the server class receives a client request, it passes it to the associated `CHerculeanSession` object for handling.

Let's examine the code for the `CHerculeanServer` class in more detail (I've omitted some of the straightforward construction code to keep the code sample as short as possible):

```
// Starts the server and constructs the shutdown object, starting the
// timer to ensure that the server will exit even if the starting client
// connection fails
void CHerculeanServer::ConstructL()
    {
    StartL(KServerName);
    iShutdown.ConstructL();
    iShutdown.Start(); // In case the client session fails to connect
    }

void CHerculeanServer::AddSession()
    {
    ++iSessionCount;
    iShutdown.Cancel();// Cancel the shutdown timer now
    }

// Decrement the session counter. Start the shutdown timer when the last
// client disconnects
void CHerculeanServer::RemoveSession()
    {
    if (--iSessionCount==0)
        iShutdown.Start();
    }

TInt CHerculeanServer::RunError(TInt aError)
    {
    if (KErrBadDescriptor==aError)
        PanicClient(Message(),EPanicBadDescriptor);
    else
        Message().Complete(aError);

    ReStart();          // Continue reading client requests
    return (KErrNone); // handled the error
    }
```

The construction of the `CHerculeanServer` class is straightforward. The shutdown timer is started when the server is first constructed, in case construction of the initial client session fails. When the first session is added successfully, the session count is incremented and the shutdown timer is canceled. The server object increments and decrements the `iSessionCount` reference count accordingly when sessions are added and removed.

`CHerculeanServer::RunError()` is called if a leave occurs in `CHerculeanSession::ServiceL()`, that is, if one of the methods which services client requests leaves. The leave code is passed to `RunError()`, which should attempt to handle the leave, returning `KErrNone` if it does so. `CServer::RunError()` was added to Symbian OS v6.0 to allow the server to manage leaves resulting from client request processing. Previously, the leaves were propagated to the active scheduler which did not have sufficient context in which to handle them.

`RunError()` panics the client if the leave code is `KErrBadDescriptor`, because this indicates that client data has been passed to the server without having been properly "descriptorized". This is indicative of a programming error, so the server is justified in panicking the client. Under all other circumstances, the server reports the error to the client by completing it using `RMessage::Complete()`. It is rarely correct to panic another thread except to indicate a programming error.

The code used by `CHerculeanServer` to panic the client is shown below. `RMessage::Panic()` uses the `RThread` handle it holds for the client thread to panic it and also completes the outstanding client request to allow the kernel to perform the necessary cleanup.

```
enum TServerPanic
    {
    EPanicBadDescriptor,
    EPanicNotSupported
    };

void PanicClient(const RMessage& aMessage,TServerPanic aPanic)
    {
    _LIT(KPanic,"HerculesServer");
    aMessage.Panic(KPanic,aPanic);
    }
```

A leave from `CHerculeanSession::ServiceL()` results in an early return from `CServer::RunL()`, which skips the call to continue requesting client messages. From Chapter 11, you'll recall that, on receipt of a client request, `CServer::RunL()` calls the `ServiceL()` method of the associated `CSharableSession`-derived object. It is for this reason that `RunError()` must call `CServer::Restart()`.

Moving on, let's consider the implementation of `CHerculean-Session`. This consists of an implementation of the `ServiceL()` method, which was declared pure virtual in the `CSharableSession` base class, and a set of private methods to handle client requests:

```
void CHerculeanSession::CreateL(const CServer& aServer)
    {// Called by the CServer framework
    CSharableSession::CreateL(aServer); // Cannot leave
    Server().AddSession();
    // Create the CAsyncHandler object (iAsyncRequestHandler)
    ...
    }

CHerculeanSession::~CHerculeanSession()
    {
    Server().RemoveSession();
    delete iAsyncRequestHandler;
    delete iClientBuf;
    }

// Handle a client request
// Leaves are handled by CHerculeanServer::RunError() which is called
// by CServer::RunL()
void CHerculeanSession::ServiceL(const RMessage& aMessage)
```

```
    {
    switch (aMessage.Function())
        {
        case ESlayNemeanLion:
            SlayNemeanLionL(aMessage);              break;
        case ESlayHydra:
            SlayHydraL(aMessage);                   break;
        case ECaptureCeryneianHind:
            CaptureCeryneianHindL(aMessage);        break;
        case ESlayErymanthianBoar:
            SlayErymanthianBoarL(aMessage);         break;
        case ECleanAugeanStables:
            CleanAugeanStablesL(aMessage);          break;
        case ECancelCleanAugeanStables:
            CancelCleanAugeanStables();             break;
        case ESlayStymphalianBirds:
            SlayStymphalianBirdsL(aMessage);        break;
        case ECancelSlayStymphalianBirds:
            CancelSlayStymphalianBirds();           break;
        case ECaptureCretanBull:  // Omitted for clarity
        case ECaptureMaresOfDiomedes:
        case EObtainGirdleOfHippolyta:
        case ECaptureOxenOfGeryon:
        case ETakeGoldenApplesOfHesperides:
        case ECaptureCerberus:
        default:
            PanicClient(aMessage, EPanicNotSupported);
            break;
        }
    }

// p[0] contains const TDesC8&
// p[1] contains TInt
void CHerculeanSession::SlayNemeanLionL(const RMessage& aMessage)
    {
    const TInt KMaxLionDes = 100;
    TBuf8<KMaxLionDes> lionDes;
    aMessage.ReadL(aMessage.Ptr0(), lionDes);
    TInt val = aMessage.Int1();
    // ... Process as necessary
    aMessage.Complete(KErrNone);
    }

// p[0] contains TPckg<THydraData>
void CHerculeanSession::SlayHydraL(const RMessage& aMessage)
    {
    THydraData hydraData;
    TPckg<THydraData> des(hydraData);
    aMessage.ReadL(aMessage.Ptr0(), des);
    // ... Process as necessary, updates hydraData.iHeadCount
    // Write hydraData update back to client
    aMessage.WriteL(aMessage.Ptr0(), des);
    aMessage.Complete(KErrNone);
    }

// p[0] contains TInt&
void CHerculeanSession::CaptureCeryneianHindL(const RMessage& aMessage)
    {
```

```
    TInt count;
    // ... Process as necessary (updates count)
    TPckgC<TInt> countDes(count);
    aMessage.WriteL(aMessage.Ptr0(), countDes);
    aMessage.Complete(KErrNone);
    }

// p[0] contains streamed CHerculesData
void CHerculeanSession::SlayErymanthianBoarL(const RMessage& aMessage)
    {
    HBufC8* desData = HBufC8::NewLC(KMaxCHerculesDataLength);
    TPtr8 readPtr(desData->Des());
    aMessage.ReadL(aMessage.Ptr0(), readPtr);
    CHerculesData* data = CHerculesData::NewLC(*desData);
    // ... Process as appropriate, passing in data
    aMessage.Complete(KErrNone);
    }

// Asynchronous method - no client parameters
void CHerculeanSession::CleanAugeanStablesL(const RMessage& aMessage)
    {// Makes an asynchronous request via the CAsyncHandler active object
    //  (initialized with aMessage to allow it to complete the client)
    ...
    }

void CHerculeanSession::CancelCleanAugeanStablesL()
    {
    ... // Calls Cancel() on the CAsyncHandler active object which
        // checks if a request is outstanding and cancels it
    }

// Asynchronous method
// p[0] contains TInt
// p[1] contains TDes8&
void CHerculeanSession::SlayStymphalianBirdsL(const RMessage& aMessage)
    {
    TInt val0 = aMessage.Int0();

    // Determine the length of the client descriptor passed to the server
    TInt clientDesMaxLen =
            aMessage.Client().GetDesMaxLength(aMessage.Ptr1());
    if (iClientBuf)
        {
        delete iClientBuf;
        iClientBuf = NULL;
        }
    // iClientBuf owned/destroyed by session
    iClientBuf = HBufC8::NewL(clientDesMaxLen);
    TPtr8 ptr(iClientBuf->Des());
    aMessage.ReadL(aMessage.Ptr1(), ptr);
    // Makes an asynchronous request via the CAsyncHandler active object
    // which is initialized with aMessage to allow it to complete the
    // client. Modifies the contents of iClientBuf and writes it back to
    // the client
    }

void CHerculeanSession::CancelSlayStymphalianBirdsL()
    {
```

```
... // Calls Cancel() on the CAsyncHandler active object which
    // checks if a request is outstanding and cancels it
}
```

`ServiceL()` consists of a `switch` statement that examines the client request opcode, using `RMessage::Function()`, and calls the associated handler method for that request. In the example code above, I've only shown some of the request handling methods that `CHerculesSession` implements.

You'll recall that the client-side request code consisted of boilerplate "packaging" of parameter data to pass to the server. By extension, the server-side request code unpacks the parameter data, performs the necessary processing upon it, repackages return data if necessary and notifies the client that the request has been completed. Let's now examine each of those stages in turn.

The parameter unpacking code is fairly routine, as you can see from the implementation of `CHerculeanSession::SlayNemeanLionL()`. The client writes a pointer to a constant `TDesC8` into the first element of the request data array. The server retrieves this data by instantiating a modifiable descriptor and using `RMessage::ReadL()`[6] to read data from the client thread into it. The `TAny` pointer to the location of the client descriptor is identified in this case by use of `RMessage::Ptr0()` – if the descriptor had been in the second slot in the request array, `RMessage::Ptr1()` would have been used, and so on.

In this example, the predetermined protocol between client and server has fixed the maximum size of the client-side descriptor as `KMaxLionDes` bytes, so the server allocates a stack-based `TBuf8` with that maximum size to receive the incoming data. However, if the size of the data is unknown at compile time, as in `SlayStymphalianBirdsL()`, the server must determine the size of the incoming descriptor to ensure that a sufficiently large descriptor is allocated on the server side to receive the client data. It can do this by calling `RThread::GetDesMaxLength()` on the client thread, passing in the pointer to the descriptor. It also needs to perform this check before writing descriptor data back to the client, to determine whether the client has allocated a large enough descriptor.

The use of a heap-based descriptor to read data from the client is more appropriate if a large or unpredictable amount of data is transferred between the client and server, because the amount of stack space available is restricted on Symbian OS.

`SlayNemeanLionL()` retrieves a `TInt` from the second element of the request data array, using `RMessage::Int1()`, which returns the client parameter in the second "slot" as an integer value. Don't let the zero-based numbering scheme confuse matters here!

---

[6] `RMessage::ReadL()` performs inter-thread data transfer by calling `RThread::ReadL()`, as discussed in Chapter 10.

Having retrieved the client parameters, the request-handling function then performs any necessary processing upon it – I've omitted this from the example code to keep it straightforward. `SlayNemeanLionL()` is a simple example because it is synchronous and doesn't package any return data to send to the client. Thus, when the request processing is finished, the server simply notifies the client by calling `RMessage::Complete()`, which signals the client thread's request semaphore to indicate request completion.

`CaptureCeryneianHindL()` shows the server writing data back to the client thread – in this case, it updates the integer value passed into the first element of the request data array. The server has an integer value, `count`, which represents the number of hinds captured. It "descriptorizes" this value using a `TPckgC` and calls `RMessage::WriteL()` to make an inter-thread data transfer into the client thread.

Earlier, I discussed in detail how the client submitted custom objects to the server, such as those of T or C classes. I described how an object of class `THydraData` was marshaled into a descriptor using the `TPckg` class, and in `CHerculesSession::SlayHydraL()` you see what happens on the other side of the client–server boundary. The server instantiates its own `THydraData` object, wraps it in a `TPckg` descriptor and then "reconstitutes" it by reading into it the descriptor passed by the client. Having done so, the server performs the necessary processing which modifies the object. It writes the changes back to the client using `RMessage::WriteL()`. In a similar manner, `CHerculesSession::SlayErymanthianBoarL()` shows how a server receives a "streamed" `CBase`-derived object in a descriptor and instantiates its own copy using the appropriate `NewLC()` method. This object can then be passed as a parameter to the appropriate internal handling function.

While most of the request handler methods shown are synchronous, `CleanAugeanStables()` and `SlayStymphalianBirdsL()` are asynchronous. The server retrieves any parameters passed from the client and passes them to an active object which is responsible for submitting requests to an asynchronous service provider and handling their completion events. To avoid complicating the code example I haven't shown the active object class here, but I discuss active objects fully in Chapters 8 and 9. The active object class must be passed a means to access the `RMessage` associated with the client request, which it will use to call `Complete()` on the client when the request has been fulfilled by the asynchronous service provider. Since it only uses the `RMessage` to complete the client, it is unnecessary for this class to hold a copy of the entire object. Commonly, the `RMessagePtr` class is used to make a copy of the client's thread handle from the `RMessage`, and the `RMessagePtr` object is then used to notify the client of the request's completion. Class `RMessagePtr` is defined in `e32std.h`.

Incidentally, a constant reference to the `RMessage` associated with the request is passed into each of the request handler methods but it may, alternatively, be retrieved by the handler methods by calling `CSharableSession::Message()`. However, the asynchronous requests *must* store a copy of the `RMessage` object, because the session may be processing another, different, request message by the time the asynchronous request completes and is handled.

12.6 Server Shutdown

The timer class which manages server shutdown is shown below:

```
const TInt KShutdownDelay=200000; // approx 2 seconds

class CShutdown : public CTimer
    {
public:
    inline CShutdown();
    inline void ConstructL();
    inline void Start();
private:
    void RunL();
    };

inline CShutdown::CShutdown()
: CTimer(-1) {CActiveScheduler::Add(this);}

inline void CShutdown::ConstructL()
{CTimer::ConstructL();}

inline void CShutdown::Start()
{After(KShutdownDelay);}

void CShutdown::RunL()
    {// Initiates server exit when the timer expires
    CActiveScheduler::Stop();
    }
```

The `CServer`-derived object owns a `CShutdown` object. As I described above, the server reference-counts its connected client sessions. The shutdown timer object is started when there are no sessions connected to the server, although it is canceled if a session connects before the timer expires. When the timeout completes, the timer's event handler calls `CActiveScheduler::Stop()` to terminate the server's wait loop and destroy the server. The timeout is used to delay shutdown and prevent excessive startup/shutdown churn caused by client connections which do not quite overlap. The server's shutdown timeout is defined by `KShutdownDelay`, which is set to 2 seconds.

12.7 Accessing the Server

Finally, for reference, here is an example of how the Hercules server may be accessed and used by a client. The client-side `RHerculesSession` class is used to connect a session to the server and wrap the caller's parameter data as appropriate, before passing it to the server.

```
void TestClientServerL()
    {
    __UHEAP_MARK; // Checks for memory leaks (see Chapter 17)
    RHerculesSession session;
    User::LeaveIfError(session.Connect());
    CleanupClosePushL(session); // Closes the session if it leaves

    _LIT8(KLionDes, "NemeanLion");
    User::LeaveIfError(session.SlayNemeanLion(KLionDes, 1));

    TVersion version(1,0,0);
    THydraData hydraData;
    hydraData.iHydraVersion = version;
    hydraData.iHeadCount = 9;
    User::LeaveIfError(session.SlayHydra(hydraData));
    ... // Checks hydraData, which was modified by the server

    TInt count;
    User::LeaveIfError(session.CaptureCeryneianHind(count));
    ... // Checks count which was set by the server

    CHerculesData* data =
            CHerculesData::NewLC(_L8("test1"), _L8("test2"), 1);
    User::LeaveIfError(session.SlayErymanthianBoar(*data));

    TRequestStatus status;
    session.CleanAugeanStables(status);
    User::WaitForRequest(status);

    // Server reads this data and updates it
    TBuf8<12> myBuf(_L8("testdata"));
    session.SlayStymphalianBirds(3, myBuf, status);
    User::WaitForRequest(status);
    ... // Inspects the contents of myBuf, modified by the server
    CleanupStack::PopAndDestroy(2, &session); // data, session
    __UHEAP_MARKEND;
    }
```

12.8 Summary

This chapter examined code for a typical client–server implementation, using a simplistic example to avoid introducing "accidental complexity". It is intended for those wishing to implement a server and its client-side access code, and to illustrate how the Symbian OS client–server architecture works in practice, reinforcing the theory described in Chapter 11.

The example is a transient server that runs in a separate process from its clients, with the client-side implementation in a separate DLL. The chapter discusses best practice in the following areas of code:

- the use of "opcodes" to identify a client request
- a typical client-side `RSessionBase`-derived class and its "boilerplate" code to submit requests to the server. The discussion included details of how to submit different types of parameter data to the server:
  - simple built-in types
  - descriptors
  - flat data (such as that contained in a `struct` or an object of a T class)
  - more complex objects, which do not have a fixed length or which contain pointers to other objects (e.g. an object of a C class).
- how to implement client-side code to start the server (which for EKA1 is different depending on whether the server is running on the Windows emulator or target hardware) and how to connect to the server
- server-side bootstrap code
- the fundamental server classes, deriving from `CServer` and `CSharableSession`, including examples of request-handling methods (for both synchronous and asynchronous requests), server-side unpacking of parameter data passed from the client, and an example of how data can be passed back to the client
- the mechanism used by a transient server to reference-count its connected client sessions and shut itself down, after a brief timeout, when all its clients have disconnected
- the implementation of a typical calling client that instantiates an object of the `RSessionBase`-derived client class and submits requests to the server.

This chapter also listed the twelve labors of Hercules, which the reader may, or may not, wish to commit to memory.

13

Binary Types

Oh Lord, forgive the misprints

Last words of Andrew Bradford, American Publisher

The executable code of any C++ component on Symbian OS is delivered as a binary package. There are two particular types discussed in this chapter: packages which are launched as a new process (`.exe`) and those that run inside an existing process, dynamically linked libraries (`.dll`). Note that I use the term "executable" in this context to refer to any binary code which may execute, as opposed to a `.exe` exclusively.

13.1 Symbian OS EXEs

On the Windows emulator Symbian OS runs within a single Win32 process. If you look at Task Manager on Windows when the emulator is running, you'll see the process, `EPOC.exe`. Within that process, each Symbian OS EXE is emulated within a separate thread.

On a phone handset running Symbian OS, commonly known as "target hardware", each EXE is launched in a separate, new process. Each process has a single main thread that invokes the sole entry point function, `E32Main()`.

On target hardware, executable code can either be built onto the phone in Read-Only Memory (ROM) when the phone is in the factory or installed on the phone at a later stage – either into the phone's internal storage or onto removable storage media such as a Memory Stick or MMC. It's a simplification, but you can generally think of ROM-based EXEs as being executed directly in-place from the ROM. This means that program code and read-only data (such as literal descriptors) are read directly from the ROM, and the component is only allocated a separate data area in RAM for its read/write data.

If an EXE is installed, rather than built into the ROM, it executes entirely from RAM and has an area allocated for program code and read-only static data, and a separate area for read/write static data. If a second copy of the EXE is launched, the program code and read-only static data area is shared, and only a new area of read/write data is allocated.

13.2 Symbian OS DLLs

Dynamic link libraries, DLLs, consist of a library of compiled C++ code that may be loaded into a running process in the context of an existing thread. On Symbian OS there are two main types of DLL: shared library DLLs and polymorphic DLLs.

A **shared library** DLL implements library code that may be used by multiple components of any type, that is, other libraries or EXEs. The filename extension of a shared library is `.dll` – examples of this type are the base user library (`EUser.dll`) and the filesystem library (`EFile.dll`). A shared library exports API functions according to a module definition (`.def`) file. It may have any number of exported functions, each of which is an entry point into the DLL. It releases a header file (`.h`) for other components to compile against, and an import library (`.lib`) to link against in order to resolve the exported functions. When executable code that uses the library runs, the Symbian OS loader loads any shared DLLs that it links to and loads any further DLLs that those DLLs require, doing this recursively until all shared code needed by the executable is loaded.

The second type of DLL, a **polymorphic** DLL, implements an abstract interface which is often defined separately, for example by a framework. It may have a `.dll` filename extension, but it often uses the extension to identify the nature of the DLL further: for example, the extension `.app` identifies an application, `.fep` a front-end processor and `.mdl` a recognizer. Polymorphic DLLs have a single entry point "gate" or "factory" function, exported at ordinal 1, which instantiates the concrete class that implements the interface. The interface functions are virtual; they are not exported and are instead accessed by the virtual function table, through a pointer to the base class interface. Polymorphic DLLs are often used to provide a range of different implementations of a single consistent interface, and are loaded dynamically at run-time by a call to `RLibrary::Load()`.

This type of DLL is often known as a "plug-in" – recognizers are a good example of plug-ins. The component that determines which plug-ins to load, instantiate and use is typically known as a framework. The framework which loads the recognizers is provided by the application architecture server (Apparc). It can load any number of recognizer plug-in DLLs, which examine the data in a file or buffer and, if they "recognize" it, return its data (MIME) type. Each recognizer plug-in exports a function at ordinal 1 that constructs and returns an instance of the `CApaDataRecognizerType` interface. The plug-in must provide a concrete class which implements the three pure virtual functions of the interface: `DoRecognizeL()`, `SupportedDataTypeL()` and `PreferredBufSize()`.

Recognizer plug-in DLLs are identified by having UID1 set to `KDynamicLibraryUid` (0x10000079), UID2 set to `KUidRecognizer` (0x10003A19) and UID3 set to a unique value to identify each individual implementation. Each recognizer has a `.mdl` file extension and its `targettype` should be MDL. Don't worry too much about this right now though – UIDs and the `targettype` specifier are described later in the chapter.

Up until Symbian OS v7.0, each framework that could be extended dynamically by plug-in code was required to take responsibility for finding the appropriate plug-ins, loading and unloading them, and calling the entry point functions to instantiate the concrete interface implementation. The ECOM framework was introduced in Symbian OS v7.0 to provide a generic means of loading plug-in code, simplifying the use of plug-ins and reducing code duplication. I'll discuss ECOM further in Chapter 14. Apparc implemented its own custom loading of recognizer plug-ins up to v8.0; in this latest release it has been modified to use ECOM.

For both types of DLL, static and polymorphic, the code section is shared. This means that, if multiple threads or processes use a DLL simultaneously, the same copy of program code is accessed at the same location in memory. Subsequently loaded processes or libraries that wish to use it are "fixed up" to use that copy by the DLL loader.

DLLs in ROM are not actually loaded into memory, but execute in place in ROM at their fixed address. DLLs running from RAM are loaded at a particular address and reference counted so they are unloaded only when no longer being used by any component. When a DLL runs from RAM,[1] the address at which the executable code is located is determined only at load time. The relocation information to navigate the code of the DLL must be retained for use in RAM. However, DLLs that execute from ROM are already fixed at an address and do not need to be relocated. Thus, to compact the DLL in order to occupy less ROM space, Symbian OS tools strip the relocation information out when a ROM is built. This does mean, however, that you cannot copy a DLL from the ROM, store it in RAM and run it from there.

On Symbian OS, the size of DLL program code is further optimized to save ROM and RAM space. In most operating systems, to load a dynamic library, the entry points of a DLL can either be identified by string-matching their name (lookup by name) or by the order in which they are exported (lookup by ordinal). Symbian OS does not offer lookup by name because this adds an overhead to the size of the DLL (storing the names of

---

[1] Loading a DLL from RAM is different from simply storing it on the internal (RAM) drive, because Symbian OS copies it into the area of RAM reserved for program code and prepares it for execution by fixing up the relocation information.

all the functions exported from the library is wasteful of space). Instead, Symbian OS only uses link by ordinal, which has significant implications for binary compatibility. Ordinals must not be changed between one release of a DLL and another, otherwise code which originally used the old DLL will not be able to locate the functions it needs in the new version of the DLL. I'll discuss binary compatibility further in Chapter 18.

13.3 Writable Static Data

While EXE components have separate data areas for program code, read-only data and writable data, DLLs do not have the latter. This has the following consequence: **Symbian OS DLLs do not support writable global data**.

So why is there no writable data section for Symbian DLLs? The reason is that any code which refers to global data must use an address to do so, rather than an offset from a pointer. When code is loaded, it must either use a fixed address to somewhere in the DLL in order to locate the data, or it must use a relocation value for the data, if it is moved to a new address. Furthermore, because DLLs are shared between processes, every process in which it loads must use the same address for the global data.[2]

Thus, each DLL that supported writable static data would need a section of RAM (a "chunk", the basic unit of system memory) allocated for it within every process that loaded it, just for static data. The smallest size of a chunk is 4 KB – which comes to a significant overhead when you consider the number of DLLs that a typical application on Symbian OS might use (often over 50), and the fact that the DLL would typically waste most of this memory, since it is unlikely to declare exactly 4 KB worth of static data.

This restriction means that you cannot use static member variables, such as those used to implement the singleton pattern (which allows only one instance of a class to be instantiated and used, and is useful for implementing a single "controller" type object). This can be inconvenient, particularly when porting code which makes uses of this idiom, or indeed any other which requires global data.

Here's an example of a simple task manager where I've included just the minimum amount of code needed to illustrate the use of a singleton.

```
// TaskManager.h // Header file
class CTask; // Defined elsewhere
```

[2] If the required address for the data has already been occupied when a DLL comes to load, the DLL will not be usable. This is quite possible, because the data is placed in a chunk which means that its address must start on a megabyte boundary, of which there are few. A workaround would be to copy the program code for the DLL, and adjust the copy to use a different address for the static data, but the overhead would be unacceptably high.

```
class CTaskManager : public CBase
    {
public:
    IMPORT_C static CTaskManager* TaskManagerL();
    IMPORT_C static void DestroyTaskManager();
public:
    IMPORT_C void AddTaskL(CTask* aTask);
    //... Omitted for clarity
private:
    CTaskManager();
// Private - destroy the singleton through DestroyTaskManager()
    ~CTaskManager();
    void ConstructL();
private:
    CTaskManager(const CTaskManager&);           // Not implemented
    CTaskManager& operator =(const CTaskManager&);// Prevents copying
private:
    static CTaskManager* iTaskManager;        // The singleton instance
private:
    RPointerArray<CTask>* iTasks;
    ...
};

// TaskManager.cpp // Implementation

// Initializes the static data
CTaskManager* CTaskManager::iTaskManager = NULL;

EXPORT_C CTaskManager* CTaskManager::TaskManagerL()
    {
    if (!iTaskManager)
        {// Construct the singleton object on first use
        CTaskManager* taskManager = new (ELeave) CTaskManager();
        CleanupStack::PushL(taskManager);
        taskManager->ConstructL();
        CleanupStack::Pop(taskManager);
        iTaskManager = iTaskManager;
        }
    return (iTaskManager);
    }

EXPORT_C void CTaskManager::DestroyTaskManager()
    {
    delete iTaskManager;
    iTaskManager = NULL;
    }

// The use of dynamic arrays is discussed in Chapter 7
 void CTaskManager::ConstructL()
    {// Creates the underlying array
    iTasks = new (ELeave) RPointerArray<CTask>(4);
    }

// Exported function through which clients manipulate the array
EXPORT_C void CTaskManager::AddTaskL(CTask* aTask)
    {
    User::LeaveIfError(iTasks->Append(aTask));
    }
```

```
CTaskManager::~CTaskManager()
    {
    if (iTasks)
        {
        iTasks->Close();
        delete iTasks;
        }
    }
```

The implementation works well in an EXE component, but because of its use of writable static data cannot be used in a DLL. If writable global data is used inadvertently, it returns an error at build time for ARM targets, emitting a message from the PETRAN tool similar to the following:

```
ERROR: Dll 'TASKMANAGER[1000C001].DLL' has uninitialised data
```

The only global data you can use with DLLs is constant global data of the built-in types, or of a class with no constructor. Thus while you may have constant global data such as this in a DLL:

```
static const TUid KUidClangerDll = { 0x1000C001 };
static const TInt KMinimumPasswordLength = 6;
```

You cannot use these:

```
static const TPoint KGlobalStartingPoint(50, 50);
// This literal type is deprecated (see Chapter 5)
static const TPtrC KDefaultInput =_L("");
static const TChar KExclamation('!');
```

The reason for this is the presence of a non-trivial class constructor, which requires the objects to be constructed at runtime. This means that, although the memory for the object is pre-allocated in code, it doesn't actually become initialized and `const` until after the constructor has run. Thus, at build time, each constitutes a non-constant global object and causes the DLL build to fail for target hardware.

Note that the following object is also non-constant because, although the data pointed to by `pClanger` is constant, the pointer itself is not constant:

```
// Writable static data!
static const TText* pClanger = (const TText*)"clanger";
```

This can be corrected as follows:

```
// pClanger is constant
static const TText* const pClanger = (const TText*)"clanger";
```

Incidentally, the issue of not allowing non-constant global data in DLLs highlights another difference between the behavior of Windows emulator builds and builds for target hardware. The emulator can use underlying Windows DLL mechanisms to provide per-process DLL data. If you do inadvertently use non-constant global data in your code, it will go undetected on emulator builds and will only fail when building for target hardware.

Symbian OS DLLs must not contain writable global or static data. The only global data which may be used are constants, either of the built-in types or of classes with no constructor.

13.4 Thread-Local Storage

As I mentioned, the lack of writable global data in DLLs can be difficult when you are porting code to run on Symbian OS. However, the operating system does provide a mechanism whereby a DLL can manage writable static data on a per-thread basis using thread-local storage, commonly known as "TLS". This allocates a single machine word of writable static data per thread for every DLL, regardless of whether the DLL uses it. Obviously, the memory overhead is far less significant than allocating a 4 KB chunk for each DLL which uses static data. However, the price of using TLS instead of direct memory access is performance; data is retrieved from TLS about 30 times slower than direct access, because the lookup involves a context switch to the kernel in order to access the data.

The use of TLS for per-thread access to global static data is safe because it avoids complications when the DLL is loaded into multiple processes. However, for writable static data to be used by multiple threads, this approach must be extended. One technique uses a server to store the data, which has the benefit of being able to use static data without the need for TLS, because it is a process. The server can make this data available to its clients, which may run in multiple threads.[3] Of course, the inter-process context switch required to access the server also has performance implications, as I discuss in Chapter 11.

The TLS slot can be used directly if you have only one machine word of data to store. For extensibility, it is more likely that you'll use it to store a pointer to a `struct` or simple T Class which encapsulates all the data you would otherwise have declared as static.

Thread-local storage is usually initialized when the DLL is attached to a thread within the DLL entry point, `E32Dll()`. Typically, code is

---

[3] You can find an example of this technique in the EpocStat product released by Peroon (***www.peroon.com/Downloads.html***), for which full source code is available for download.

added to construct the `struct` containing the global data and store it in thread-local storage using the static function `Dll::SetTLS()`. (When the DLL is detached from the thread, the TLS slot should be reset and the static data deleted.) To access the data, you should use the static function `Dll::Tls()`. This will return a `TAny*` which can be cast and used to access the data. For simplicity, you may wish to provide a utility function, or set of functions, to access the data from a single point.

Here's some example code to illustrate the use of thread-local storage when implementing a task manager which runs as a single instance. The code is a modification of the previous version above, and can now be used within a DLL:

```
// TaskManager.h
class CTask; // Defined elsewhere

class CTaskManager : public CBase
    {
public:
    IMPORT_C static CTaskManager* New();
    ~CTaskManager();
public:
    IMPORT_C void AddTaskL(CTask* aTask);
    // ... omitted for clarity
private:
    CTaskManager();
    void ConstructL();
private:
    CTaskManager(const CTaskManager&);              // Not implemented
    CTaskManager& operator =(const CTaskManager&); // prevents copying
private:
    RPointerArray<CTask>* iTasks; // Single instance
    };

// Accesses the task manager transparently through TLS
inline CTaskManager* GetTaskManager()
    { return (static_cast<CTaskManager*>(Dll::Tls())); }

// TaskManager.cpp
GLDEF_C TInt E32Dll(TDllReason aReason)
    {
    TInt r =KErrNone;
    CTaskManager* taskManager = NULL;
    switch (aReason)
        {
#ifdef __WINS__
   //   On Windows, DLL attaches to the current process, not a thread
        case EDllProcessAttach:
#else
        case EDllThreadAttach:
#endif
        // Initialize TLS
        taskManager = CTaskManager::New();
        if (taskManager)
            {
            Dll::SetTls(taskManager);
```

```
            }
        break;
#ifdef __WINS__
        case EDllProcessDetach:
#else
        case EDllThreadDetach:
#endif
        // Release TLS
        taskManager = static_cast<CTaskManager*>(Dll::Tls());
        if (taskManager)
            {
            delete taskManager;
            Dll::SetTls(NULL);
            }
        break;
        default:
        break;
        }

    return(r);
    }

// Non-leaving because it is called by E32Dll() which cannot leave
EXPORT_C CTaskManager* CTaskManager::New()
    {
    CTaskManager* me = new CTaskManager();
    if (me)
        {
        me->iTasks = new RPointerArray<CTask>(4);
        if (!me->iTasks)
            {
            delete me;
            me = NULL;
            }
        }
    return (me);
    }
```

If you look at the documentation for class DLL in your SDK, you may find that it directs you to link against `EUser.lib` to use the `Tls()` and `SetTls()` functions. You'll find this works for emulator builds, but fails to link for ARM. This is because the methods are not implemented in `EUser.dll` – you should now link against `edllstub.lib`, before linking to `EUser.lib`.

Thread-local storage (TLS) can be used to work around the prohibition of writable global data. However, the use of TLS affects performance; data is retrieved from TLS about 30 times slower than direct access because the lookup involves a context switch to the kernel.

13.5 The DLL Loader

An interesting scenario can arise when you attempt to replace a DLL with a version you have rebuilt, perhaps a debug version or one that includes tracing statements to help you track down a problem. You may find that it is not "picked up" by the loader, which continues to run the original version.

As I described earlier, a DLL running from RAM is only loaded once, even if multiple processes use it. The DLL loader uses reference counting, and only unloads the DLL when none of its clients are running. So, if the DLL is loaded by other processes, it will not be unloaded and cannot be replaced by a newer version until the original is unloaded. This is also relevant if you are attempting to replace a ROM-based[4] DLL. If a ROM version of the DLL you wish to replace is used before your component loads, your component will also end up using the ROM-based version. In effect, this means that you can never replace a DLL which is used by the application launcher shell.

The DLL is loaded by name lookup (with additional UID checking which I'll describe shortly). If a DLL is not already loaded, the DLL loader uses a particular lookup order to find it, as follows:

1. The same directory as the process wishing to load the DLL (often `c:\system\libs`).
2. The directory associated with the filesystem's default session path (which is usually the root of the `C:` drive).
3. The `\system\libs` directories on all available drives, in the following order: `C:` (the internal storage drive), `A:`, `B:`, `D:`, `E:`, `...`, `Y:` and finally, `Z:` (the ROM).

If you wish to replace a DLL, you should ensure that you put the replacement where it will be loaded before the original version.

13.6 UIDs

A UID is a signed 32-bit value which is used as a globally **u**nique **id**entifier. Symbian OS uses a combination of up to three UIDs to create a `TUidType` compound identifier. UIDs are used extensively, for example, to identify the type of a component and to verify that it is compatible and supports a particular interface. Thus, a DLL can register a *type* to reflect the interface it is implementing. The DLL loader can check the type of

[4] In case you were wondering, you can never replace a DLL used by a component on ROM, because the ROM binaries are linked together when the ROM is built. However, you can replace a ROM DLL if the component that is using it isn't running from ROM.

a DLL (using `RLibrary::Type()`) to determine whether a component is of the correct type, and prevent other files which may share the same name from being loaded.

The three UIDs are identified as UID1, UID2 and UID3 and are generally used as follows:

- UID1 is a system-level identifier to distinguish between EXEs (`KExecutableImageUid = 0x1000007a`) and DLLs (`KDynamicLibraryUid = 0x10000079`)
- UID2 distinguishes between components having the same UID1, for example between shared libraries (`KSharedLibraryUid = 0x1000008d`) or polymorphic DLLs such as applications (`KUidApp = 0x100039CE`), recognizers (`0x10003A19`) or front-end processors (`0x10005e32`)
- UID3 identifies a component uniquely. In order to ensure that each binary that needs a distinguishing UID is assigned a genuinely unique value, Symbian manages UID allocation through a central database.[5]

For test code, or while your code is under development, you may prefer to use a temporary UID from a range reserved for development only. These values lie in the range 0x01000000 to 0x0fffffff. You must still take care to avoid re-using UIDs in this region because a UID clash may prevent a library from loading. For this reason, these values must not be used in any released products.

You don't need to specify UID1 for a component, because it is defined implicitly by the `targettype` you choose for your component (I'll discuss the different options of `targettype` in the next section). For example, a component which is specified as `targettype epocexe` is assigned `UID1=KExecutableImageUid` by the system, which is built directly into the binary. By comparison, `targettype dll` (for a shared library component) is automatically assigned `UID1=KDynamicLibraryUid`. A component's second and third UIDs, if used, must be specified as hexadecimal values in the `.mmp` file of the component.

For native binaries, Symbian OS uses UIDs as the primary means of identification, rather than filenames or filename extensions.

[5] You can make a request to Symbian for allocation of one or more UIDs by submitting an email with the subject "UID Request" to Symbian (***UID@symbiandevnet.com***). You can ask to be assigned a single UID or a block of values, usually no more than ten, although you will be granted more if you state your reasons for needing them. Your submission should also include your name (or that of the application) and your return email address.

13.7 The targettype Specifier

The targettype specifier in the .mmp (project) file allows you to define the particular binary type of your component. The targettype is not necessarily the extension assigned to the component when it builds, although it may be, but categorizes it for the build tools. I describe below the most common binary types you'll encounter on Symbian OS. Various other plug-in targettypes, such as app, fep, mdl, prn and ecomiic, may be used for a polymorphic DLL.

targettype epocexe

You would logically think that any component running on Symbian OS as a separate, "out-of-process" component, such as a server, would be built with the .exe extension. However, as I described earlier, the Windows emulator runs in a single process. So, while you can run multiple Symbian OS processes on hardware (ARM builds), on Windows each Symbian OS process is built as a DLL which runs inside a separate thread that emulates a Symbian OS process within the single Win32 emulator process, EPOC.exe.

On target hardware, if you browse the file system, select and click on a .exe file, it will start a different process. However, this is not possible on the emulator, which is why the epocexe type was introduced in v6.0 to simulate this behavior. It instructs the build tools to build the component as a .exe for multi-process hardware platforms and as a .dll for single-process emulator builds. This allows an epocexe component to be launched directly, both in the single process of the emulator and as a separate process on target hardware. An example of a typical epocexe component is the contacts server (built as cntsrv.dll on Windows and cntsrv.exe for target hardware).

This is true for versions of Symbian OS earlier than v8.0. The new kernel in Symbian OS v8.0 has more complete process emulation on Windows, and an EXE may now be launched directly both in the emulator (although it still runs within the single emulator process) and on hardware. As a result, targettype epocexe is no longer needed and code which runs as a separate process, such as a Symbian OS server, may now be built as an EXE for both Windows and hardware platforms.

Components of this targettype should implement WinsMain(), which is exported as ordinal 1 for emulator builds, to form the DLL entry point. There should be no other exports besides this entry point for emulator builds, and there need be no exported functions at all for ARM builds. For example:

```
GLDEF_C TInt E32Main() // Process entry point function
    {
```

```
    ... // Omitted for clarity
    return (KErrNone);
    }

#if defined(__WINS__)
EXPORT_C TInt WinsMain()
    {
    E32Main();
    return (KErrNone);
    }
TInt E32Dll(TDllReason)
    { // DLL entry point for the DLL loader
    return (KErrNone);
    }
#endif
```

targettype exedll

This targettype exists as an alternative to epocexe and allows separate process components to export functions to clients for both hardware and emulator builds. Like epocexe, the build tools interpret this differently on each platform and build the component as .exe for the multi-process hardware (ARM) platforms and as .dll for single-process emulator platforms. An example of this targettype is the random server, which builds as randsvr.exe for ARM builds and randsvr.dll to run on the emulator.

A component of this targettype must implement the DLL entry point function E32Dll() for emulator builds only, to allow it to be loaded as a DLL. This should be the first exported function in the .def file.

In releases up to and including v5.0, the epocexe type did not exist and exedll was used instead. This targettype is also due to be retired in EKA2[6] versions of Symbian OS, because the enhanced process emulation, described above, allows out-of-process components to be built as a .exe for both ARM and emulator platforms. However, to allow this type of component to export functions, a new targettype will be introduced to replace it. This will be called exexp and, on all platforms, will build components as .exe, which may export any number of entry point functions.

targettype exe

The build tools build a component of this targettype to have the .exe extension on both the emulator and target hardware. On EKA1, it is only used for basic console applications such as Symbian OS command-line

[6] You may recall from Chapter 10 that Symbian identifies the new hard real-time kernel in Symbian OS v8.0 as 'EKA2' which stands for 'EPOC Kernel Architecture 2'. The kernel in previous versions of Symbian OS is referred to as EKA1.

("Text Shell") test code, which I discuss further in Chapter 17. Text Shell programs use the text window server and the programs are launched by having integral emulator support. On EKA1 releases of Symbian OS, you can only run them on the Windows emulator by launching them directly from the command prompt on the PC, by running them from the debugger or by launching the text shell, `EShell.exe`, from the command line of the PC and then invoking your test executable from inside it. On EKA2, Symbian OS process emulation has been enhanced on Windows, so you can directly load the EXE from the command line, as previously, but you can also start it from within the emulator by selecting it from the application launcher shell, file manager or any other application which launches processes. On EKA2, the Windows emulator corresponds more closely to behavior on hardware where, on all releases of Symbian OS, an EXE may be invoked directly.

targettype lib

This `targettype` is used for a static library, which is a file to which other executable code links to resolve references to exported functions. The component will build with a `.lib` extension.

13.8 Summary

This chapter examined the nature of DLLs and EXEs on Symbian OS. It described how Symbian OS EXEs are emulated on Windows, and described the difference between running an EXE from ROM and when installed to internal storage or removable media, on hardware.

Symbian OS has two types of dynamic link library: shared library and polymorphic DLL. All Symbian OS DLLs built into the ROM are stripped of relocation information to minimize their size. Additionally, all Symbian OS code links to DLLs by ordinal rather than by name, which reduces the amount of space required in the DLL export table. The chapter also gave brief details of how DLLs load, including the basic details of the DLL loader.

Symbian OS UIDs are used to identify components by type and give binaries a unique identity. The relationship between UID and `targettype` (`epocexe`, `exedll`, `exexp`, `exe`, `dll`, `lib` and polymorphic DLL types such as `app` or `fep`) was discussed. Each `targettype` was explained in terms of its binary type on hardware and emulator platforms, and any differences occurring between EKA1 and EKA2 releases of Symbian OS.

The chapter also examined the reasons why no Symbian OS DLL may have modifiable static or global data, and described how thread-local storage can be used instead to provide access to global data within a single thread. It described why the use of thread-local storage can have performance disadvantages but can be useful when porting code which previously relied on static data, for example, through use of the singleton pattern.

14

ECOM

Go on, get out! Last words are for fools who haven't said enough!
Karl Marx

In Chapter 13, I described polymorphic "plug-in" DLLs which implement an abstract interface and are dynamically loaded at run-time to extend a "framework" process. There are three roles associated with an extensible plug-in framework:

- an interface, which defines a service
- an implementation of that interface (within a polymorphic plug-in DLL); the specifics of an implementation may not be known until run-time and several implementations of the interface may exist
- a framework, which provides clients with access to all the implementations of a specific interface, including a way to determine which implementations are available (resolution), and to manage the creation and destruction thereof.

Prior to Symbian OS v7.0, each framework used its own custom code to identify the plug-in type and the nature of the specific implementation and to load and unload individual plug-ins as appropriate. To simplify such framework code, ECOM was added to Symbian OS v7.0.

14.1 ECOM Architecture

ECOM is a generic and extensible framework by which abstract interfaces can be defined and their implementations identified, loaded and managed. Frameworks can delegate their plug-in identification and instantiation to ECOM, and do not have to duplicate complex code which does not bear any direct relevance to the required behavior of the framework itself. The architecture:

- Identifies all the concrete implementations of a particular interface.

- Allows a client of that interface to specify dynamically which interface implementation should be used. The selection process is called **resolution**. ECOM provides a default resolver but interface definitions can provide their own specialized resolvers where required.
- Instantiates an instance of the concrete class which implements that interface by calling the appropriate factory function.

The class is implemented in an ECOM plug-in DLL and ECOM loads it into the process in which the caller is running. The plug-in may be loaded by a number of different clients running simultaneously in different processes. ECOM uses reference counting to manage this, and unloads it when all its clients have released it.

The ECOM architecture is used transparently by interface clients without the need for them to call any ECOM-specific functions. ECOM creates an instance of an interface implementation dynamically, either by selecting a default implementation or by using a cue from the client as a way of identifying the particular implementation required. The means by which ECOM identifies and loads the correct plug-in is hidden from the client. Once an object of the concrete class is instantiated, the caller accesses it transparently through the interface. The client does not need to access the concrete class itself, nor does it need to access ECOM directly, although it must link against the ECOM client library (`ECOM.lib`).

ECOM itself uses a client–server architecture, which is discussed in more detail in Chapters 11 and 12. The ECOM client class, `REComSession`, provides functions to list, instantiate and destroy interface implementations. A single instance of the ECOM client session exists per thread, and is accessed by interface instantiation and destruction code, which uses a set of static functions.

The ECOM server manages requests to instantiate concrete instances of an interface. By using resource file metadata provided by each ECOM plug-in, it constructs a registry of all interface implementations installed on the device. The server constantly monitors which implementations are available by means of filesystem scanning, which determines when plug-ins have been added or removed, by installation or on removable media.

An interface implemented by an ECOM plug-in has two characteristic features. Firstly, it is an abstract class that defines a set of one or more pure virtual functions. This is the standard definition of an interface. However, in addition, the interface must also provide one or more factory functions to allow clients to instantiate an interface implementation object.

The factory functions do not instantiate the object directly, because an interface cannot predict which classes will implement it. Instead, the factory functions issue requests to the ECOM framework which, at runtime, instantiates the appropriate implementation class dynamically. As I described above, in order to determine which implementation to instantiate, ECOM is given a cue, which may be a UID or some text which is passed to a resolver. Alternatively, a list of all implementations of a particular interface can be returned to the caller, which can then specify the one to use. I'll illustrate both methods of instantiation in code later in the chapter.

If the only contact you intend to make with ECOM is as an ECOM interface client, the good news is that you don't need to read any more of this chapter. However, you may find it useful to go to the end, where you'll find some example code for a typical interface client and a brief summary of the contents of the rest of this chapter.

This chapter discusses how to define an interface and implement it in an ECOM plug-in DLL. I'll illustrate the main points with example code that defines an abstract base class interface, `CCryptoInterface`, which can be used to encrypt and decrypt the contents of an 8-bit descriptor.[1]

I'll show how ECOM can be used to select dynamically which of two possible concrete implementations of the interface to use. In the example, I've assumed that there are just two implementations of the interface, which differ because one uses a cryptography library implemented in software while the other uses a cryptographic hardware module, accessed by means of a device driver. Of course, numerous implementations of this example interface could exist, provided by separate plug-in DLLs, for various cryptography libraries ported to Symbian OS. However, for simplicity, this example implements both concrete classes in a single plug-in DLL.

Figure 14.1 shows the relationship between the client, interface and implementation classes and ECOM, using the classes from the example code for illustration.

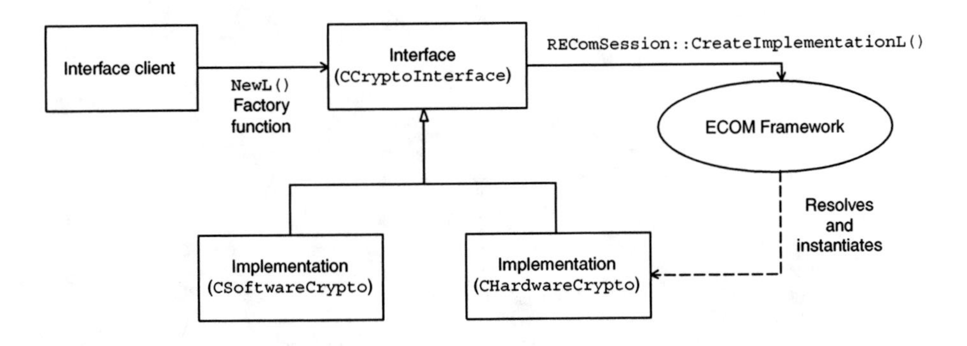

Figure 14.1 Relationship between client, interface and implementation classes and ECOM

[1] I invented the fictional `CCryptoInterface` class, and all the sample code in this chapter, merely as an example of how to use ECOM. Any resemblance to Symbian OS cryptography libraries, living or dead, is purely coincidental.

14.2 Features of an ECOM Interface

Let's start by considering the features of an ECOM interface:

- As expected, the interface will define a set of pure virtual functions which a concrete instance will implement.
- In addition, the interface must also provide one or more factory functions that pass a cue to ECOM to enable it to instantiate an object of the correct implementation (in the example below, the interface definition has two static `NewL()` functions).
- The interface must also provide a means for its clients to release it, such as a destructor to allow it to be deleted, or a method such as `Release()` or `Close()`.
- An ECOM interface definition must also have a `TUid` data member which is used internally by ECOM to identify an implementation instance for cleanup purposes.

Here is the definition of the example interface class, `CCrypto-Interface`. You'll notice that the example interface class is a C class rather than an M class, which is what you may automatically have expected for an abstract interface definition from the discussion of the standard Symbian OS class types in Chapter 1. The reason for this is clear on further inspection, because an ECOM interface class has features which are atypical of an M class, such as the static instantiation method and a `TUid` data member.

```
class CCryptoInterface : public CBase
    {
public:
    enum TAlgorithm { EDES, E3DES, EAES, ERC2, ERC4 };

    // Instantiation of a default object of this type
    IMPORT_C static CCryptoInterface* NewL();

    // Instantiation using a cue to identify the implementation to use
    IMPORT_C static CCryptoInterface* NewL(const TDesC8& aCue);

    IMPORT_C virtual ~CCryptoInterface();

    // List all implementations of this interface
    IMPORT_C static void
            ListImplementationsL(RImplInfoPtrArray& aImplInfoArray);
public:
    // Interface functions to be implemented by a concrete instance
    virtual void EncryptL(TDesC8& aPlaintext, TDesC8& aKey,
            TDes8& aCiphertext, CryptoInterface::TAlgorithm) = 0;
```

```
    virtual void DecryptL(TDesC8& aCiphertext, TDesC8& aKey,
            TDes8& aPlaintext, CryptoInterface::TAlgorithm) = 0;
private:
    TUid iDtor_ID_Key; // Identification on cleanup
    };
```

This example interface derives directly from `CBase`, but it is not mandatory for an ECOM interface definition to do so. Interfaces may, for example, equally well derive from another C class such as `CActive`, which, of course, itself derives from `CBase`. Although the interface is defined as a C class, you'll notice that a couple of the usual characteristics of a C class are missing. Firstly, the class has no explicitly declared constructor because it is an abstract base class and will not itself be constructed.[2] In addition, although the class defines a static `NewL()` factory function, it does not define a `ConstructL()` method for second-phase construction. This is because the `NewL()` factory method of `CCryptoInterface` does not use standard two-phase construction, but instead calls ECOM to instantiate a suitable implementation object.

An ECOM interface has the following characteristics:

- **It is an abstract class with a set of one or more pure virtual functions.**
- **It must provide one or more factory functions to allow clients to instantiate an interface implementation object.**
- **It must provide a means for its clients to release it, such as a destructor or a method such as `Release()` or `Close()`.**
- **It has a `TUid` data member, used internally to identify an implementation instance for cleanup purposes.**

14.3 Factory Methods

Let's look at the implementation of the two static factory methods in more detail. The overload of `NewL()` that takes no parameters creates an object of the default implementation of `CCryptoInterface`. In the example

[2] The compiler generates an implicit default constructor when instances of classes deriving from `CCryptoInterface` are created, in order to construct the virtual function table for the class.

code below, the method hard-codes a specific implementation (CSoftwareCrypto) by passing ECOM its UID.[3] This method of specifying a particular class assumes that the UID of at least one implementation class is known when the factory function for the interface is compiled.

An alternative, more loosely-coupled, technique for instantiating a default implementation uses the ECOM resolver. The resolver can be passed a cue to enable it to instantiate, for example, the first concrete class it discovers which implements that interface. I'll illustrate how to do this later, when I discuss how to customize the ECOM resolver.

```
EXPORT_C CCryptoInterface* CCryptoInterface::NewL()
    {
    // Hard-coded to use CSoftwareCrypto by default
    const TUid KSWCryptoUid = { 0x10008EE5 };

    TAny* defaultCrypto =REComSession::CreateImplementationL(KSWCryptoUid,
            _FOFF(CCryptoInterface, iDtor_ID_Key));

    return (reinterpret_cast<CCryptoInterface*>(defaultCrypto));
    }
```

REComSession::CreateImplementationL() is passed the UID of the required implementation and the offset[4] of the iDtor_ID_Key member of the interface class. Don't worry too much about this ECOM-internal housekeeping, which initializes iDtor_ID_Key so ECOM can later identify the object for cleanup. CreateImplementationL() returns a TAny* which must be re-cast to the interface pointer before returning it to the caller. There are a number of other overloads of REComSession::CreateImplementationL(), which you'll find documented in detail in your preferred SDK (from Symbian OS v7.0 onwards).

The second NewL() factory method takes a cue parameter, which it passes to the default ECOM resolver. The resolver uses it to determine the concrete instance of the implementation to be instantiated, by matching the cue against the default_data attributes for registered implementations of that interface. I'll discuss the role of the default resolver and ECOM resource files in more detail later.

```
// interface_uid for CCryptoInterface
const TUid KCCryptoInterfaceUid = {0x10008EE0};
```

---

[3] Implementations are identified by their implementation_uid attribute, which is declared in the resource file associated with the plug-in DLL, as I'll discuss later in the chapter.

[4] The _FOFF macro is defined as follows in e32def.h:

```
#define _FOFF(c,f) ((TInt)(&((c *)0)->f))
```

It returns the offset in bytes of a member variable f of class c

```
EXPORT_C CCryptoInterface* CCryptoInterface::NewL(const TDesC8& aCue)
    {// Resolution using the default ECOM resolver
    TEComResolverParams resolverParams;
    resolverParams.SetDataType(aCue);
    // Allows wildcards in the string match
    resolverParams.SetWildcardMatch(ETrue);

    TAny* cryptoInterface =
            REComSession::CreateImplementationL(KCCryptoInterfaceUid,
            _FOFF(CCryptoInterface, iDtor_ID_Key), NULL,
            resolverParams));

    return (reinterpret_cast<CCryptoInterface*>(cryptoInterface));
    }
```

Besides the `TEComResolverParams` object, which encapsulates the resolver cue, other parameters passed to `REComSession::CreateImplementationL()` include the UID of the requested interface and the offset of `iDtor_ID_Key` as discussed previously. The third parameter, NULL in the example above, can be used to pass data to the initialization method of the concrete class. If `CreateImplementationL()` cannot find an appropriate interface implementation for the cue given, it leaves with `KErrNotFound`.

By calling `NewL()`, an interface client instantiates an object which implements that interface. The implementation type is either determined by the cue passed into `NewL()` or is the default type. Because the caller uses the interface to perform instantiation, it does not need information about the class that actually implements that interface. However, `REComSession` does provide a function, `ListImplementationsL()`, which returns information about all the implementations of a given interface in an object of type `RImplInfoPtrArray`. This can be used to provide interface clients with a list of every implementation class available. I'll illustrate how a client might use this function at the end of this chapter.

```
EXPORT_C void CCryptoInterface::ListImplementationsL(RImplInfoPtrArray&
aImplInfoArray)
    {
    REComSession::ListImplementationsL(KCCryptoInterfaceUid,
            aImplInfoArray);
    }
```

The `CCryptoInterface` class also defines a virtual destructor, which calls the ECOM framework, passing `iDtor_ID_Key` to identify the object. This enables ECOM to perform reference counting and cleanup as necessary. Subclasses will inherit this destructor code, which C++ calls after the destructor of the derived class.

```
EXPORT_C CCryptoInterface::~CCryptoInterface()
    {// Notify ECOM that this object is being deleted
    REComSession::DestroyedImplementation(iDtor_ID_Key);
    }
```

14.4 Implementing an ECOM Interface

Having discussed `CCryptoInterface`, let's move on to consider how it is implemented. Concrete classes must be defined in an ECOM plug-in DLL, which is built with `targettype ECOMIIC` (which stands for "ECOM Interface Implementation Collection"). Here's the `.mmp` file for the example plug-in DLL:

```
// ECryptoExample.mmp

TARGET ECryptoExample.dll
TARGETTYPE ECOMIIC

// UID2 = ECOM plug-in DLL recognition, UID3 = unique UID for this DLL
UID 0x10009D8D 0x10009EE1

SOURCEPATH .
SOURCE // ... Omitted for clarity
USERINCLUDE // ... Omitted for clarity
SYSTEMINCLUDE \epoc32\include \epoc32\include\ecom

RESOURCE 10009EE1.RSS // The resource file for this implementation

LIBRARY ECOM.lib // ECOM plug-in DLLs must link against ECOM.lib
LIBRARY ...      // Other libraries as required
```

The plug-in is a polymorphic Symbian OS DLL and must define a standard DLL entry point as follows:

```
TBool E32Dll()
    {
    return (ETrue);
    }
```

Within a plug-in DLL, an ECOM interface collection may contain one or more implementations of one or more ECOM interfaces and/or multiple implementations of the same interface. The sample code implements two subclasses of `CCryptoInterface`.

```
// implementation_uid = 0x10008EE5 (discussed later)
class CSoftwareCrypto : public CCryptoInterface
    {
```

```
public:
    static CSoftwareCrypto* NewL();
    ~CSoftwareCrypto();
    // Implementation using a software cryptography library
    virtual void EncryptL(TDesC8& aPlaintext, TDesC8& aKey,
            TDes8& aCiphertext);
    virtual void DecryptL(TDesC8& aCiphertext, TDesC8& aKey,
            TDes8& aPlaintext);
private:
    CSoftwareCrypto();
    void ConstructL();
private:
    ... // Omitted for clarity
    };

// implementation_uid = 0x10008EE4 (discussed later)

class CHardwareCrypto : public CCryptoInterface
    {
public:
    static CHardwareCrypto* NewL();
    ~CHardwareCrypto();
    // Implementation using hardware & device drivers
    virtual void EncryptL(TDesC8& aPlaintext, TDesC8& aKey,
            TDes8& aCiphertext);
    virtual void DecryptL(TDesC8& aCiphertext, TDesC8& aKey,
            TDes8& aPlaintext);
private:
    CHardwareCrypto();
    void ConstructL();
private:
    ... // Omitted for clarity
    };
```

Besides the interface functions, every implementation must have an instantiation function which it "registers" with ECOM. The classes above each define a static `NewL()` factory function, which uses two-phase construction as described in Chapter 4.

ECOM calls the `NewL()` factory function of a class when a call to `REComSession::CreateImplementationL()` specifies that class directly or a given cue resolves to that implementation.

A plug-in DLL "registers" these instantiation functions with ECOM by exporting a standard function (`ImplementationGroupProxy()`) which returns a pointer to an array of `TImplementationProxy` objects. This is the only function that a polymorphic ECOM plug-in DLL exports. Each `TImplementationProxy` object represents a single implementation class and contains the `TUid` which identifies the implementation (this should match the `implementation_uid` value for the class in its registration resource file) and a pointer to its instantiation method.

```
#include <ImplementationProxy.h> // ECOM header file
#include "CHardwareCrypto.h"     // Class definition for CHardwareCrypto
#include "CSoftwareCrypto.h"     // Class definition for CSoftwareCrypto
```

```
// An array of TImplementationProxy objects which connect each
// implementation with its instantiation function
const TImplementationProxy ImplementationTable[] =
    {
    {{0x10008EE4}, CHardwareCrypto::NewL},
    {{0x10008EE5}, CSoftwareCrypto::NewL}
    };

// Exported proxy function to resolve instantiation methods for an ECOM
// plug-in DLL
EXPORT_C const TImplementationProxy* ImplementationGroupProxy(TInt&
      aTableCount)
    {
    aTableCount = sizeof(ImplementationTable) /
          sizeof(TImplementationProxy);
    return ImplementationTable;
    }
```

Slightly confusingly, ECOM uses a number of different UIDs for identification:

- **`interface_uid` is used to identify a specific interface**
- **`implementation_uid` is used to identify each concrete class that implements an interface**
- **an ECOM plug-in DLL, which can contain one or more ECOM interface implementations, is identified using UID3 for the DLL (as described in Chapter 13).**

14.5 Resource Files

Earlier, I briefly mentioned that ECOM uses resource files to build up a "registry" of all the ECOM plug-ins installed. Let's now examine the detail. Each plug-in DLL must supply a compiled `.rss` resource, of a particular format, in order to register with the ECOM framework. Both the plug-in DLL and the resource file should be built into the `\system\libs\plugins\` directory.

The resource file lists all the plug-in properties, including information about each concrete class implementing an ECOM interface. ECOM associates the resource file with the plug-in DLL by name – it should be named using the hexadecimal value of the third UID of the plug-in DLL (for the example below, this is `10008EE1.rss`).

```
// 10008EE1.RSS
// Registry file for the CCryptoInterface Implementation Collection

#include "RegistryInfo.rh" // Defines the resource structures

RESOURCE REGISTRY_INFO theInfo
    {
    // UID3 for the plug-in; must match the name of this file
    dll_uid = 0x10008EE1;

    interfaces =  // interfaces info
    {
    INTERFACE_INFO
        {
        interface_uid = 0x10008EE0; // UID of CCryptoInterface

        implementations =
            {
            IMPLEMENTATION_INFO // Info for CHardwareCrypto
                {
                // Identifies the specific implementation
                implementation_uid = 0x10008EE4;
                version_no = 1;
                display_name = "Hardware Cryptography";
                // Used for cue lookup by the default ECOM resolver
                default_data = "HW";
                opaque_data = "";
                },

            IMPLEMENTATION_INFO // Info for CSoftwareCrypto
                {
                // Identifies the specific implementation
                implementation_uid = 0x10008EE5;
                version_no = 1;
                display_name = "Software Cryptography";
                // Used for cue lookup by the default ECOM resolver
                default_data = "SW";
                opaque_data = "";
                }
            };
        }
    };
}
```

A single `REGISTRY_INFO` structure is used to declare all the implementations of ECOM interfaces (identified by `interface_uid`) available in the plug-in DLL (identified by `dll_uid`). The details of each implementation are declared inside separate `IMPLEMENTATION_INFO` structures. The default ECOM resolver uses the `default_data` attribute to resolve a cue passed to a factory instantiation function. In addition, it is possible to customize the resolver to use the `opaque_data` attribute for lookup. Customized resolution can be useful to extend the selection criteria from those of the default ECOM resolver, for example by implementing case-insensitive lookup.

A customized resolver is actually an implementation of an ECOM interface, `CResolver`, and it should thus be implemented in an ECOM plug-in DLL and registered with ECOM by supplying a resource file as described above. The `interface_uid` for `CResolver` is 0x10009D0 – you should always use this value in the resource file created for your custom resolver.

Such a resolver must implement the pure virtual functions `IdentifyImplementationL()` and `ListAllL()` declared in `CResolver`. It should also specify a factory creation function which takes an `MPublicRegistry` reference parameter; this object gives the resolver access to a list of implementations of a specified interface.

`IdentifyImplementationL()` should be implemented to identify the most appropriate implementation of a specified interface, according to a cue. It will use the `MPublicRegistry` object to obtain information about each implementation. `ListAllL()` must return a list of all implementations which match the cue.

```
class CCryptoResolver : public CResolver
    {
public:
    // Factory function
    static CCryptoResolver* NewL(MPublicRegistry& aRegistry);
    ~CCryptoResolver();
public: // From CResolver (see resolver.h)
    /** Request that the resolver identify the most appropriate
        interface implementation, returning the UID or KNullUid if no
        match is found */

    TUid IdentifyImplementationL(TUid aInterfaceUid,
            const TEComResolverParams& aAdditionalParameters) const;

    /** Return a pointer to an array of all the implementations which
        satisfy the specified interface */

    RImplInfoArray* ListAllL(TUid aInterfaceUid,
            const TEComResolverParams& aAdditionalParameters) const;

    ... // Omitted
    };
```

The custom resolver may be used by the interface instantiation functions; for example, the `CCryptoInterface::NewL()` factory method may be extended to use a custom resolver as follows:

```
CCryptoInterface* CCryptoInterface::NewL()
    {// The implementation instantiated is the first found by resolver
    _LIT8(KAny,"*");
    TEComResolverParams resolverParams;
    resolverParams.SetDataType(KAny());
    resolverParams.SetWildcardMatch(ETrue);
```

```
// UID of the crypto resolver
const TUid KCustomResolverUid = {0x10008EE6};

TAny* cryptoInterface =
        REComSession::CreateImplementationL(KCCryptoInterfaceUid,
        _FOFF(CCryptoInterface,iDtor_ID_Key), NULL, resolverParams,
         KCustomResolverUid));

return (reinterpret_cast<CCryptoInterface*>(cryptoInterface));
}
```

The third UID of an ECOM plug-in DLL is used to name its associated compiled resource, which contains its ECOM "registration" information.

14.6 Example Client Code

So how does a client use an ECOM plug-in? As I have already described, the caller doesn't need to be aware of the details and simply uses the factory function supplied by the interface, supplying a cue if it is required. ECOM takes care of the details of locating and instantiating the appropriate implementation. An interface client must simply include the header files which define the interface, link against `ECOM.lib` and use the factory instantiation functions as appropriate:

```
#include "CryptoInterface.h"

void GetDefaultCryptoL()
    {// Get the default implementation of CCryptoInterface
    CCryptoInterface* crypto = CCryptoInterface::NewL();
    CleanupStack::PushL(crypto);
    ...
    CleanupStack::PopAndDestroy(crypto);
    }

void GetSpecifiedCryptoL()
    {// Use a cue - gets CCryptoInterface implementation which uses
      // hardware support
    _LIT8(KHardware,"HW");
    CCryptoInterface* crypto = CCryptoInterface::NewL(KHardware);
    CleanupStack::PushL(crypto);
    ...
    CleanupStack::PopAndDestroy(crypto);
    }

void GetAllCryptoL()
    {
// Get all implementations using CCryptoInterface::ListImplementationsL()
    RImplInfoPtrArray infoArray;
```

```
CCryptoInterface::ListImplementationsL(infoArray);
// infoArray is not leave-safe, but use of the cleanup stack is
// omitted for clarity. See Chapter 3 for more information.
...

CCryptoInterface* crypto;
for (TInt i =0; i< infoArray.Count(); i++)
    {
    // Retrieves default_data for each
    TPtrC8 dataType = infoArray[i]->DataType();
    crypto = CCryptoInterface::NewL(dataType); // Use this as a cue
    CleanupStack::PushL(crypto);
    ...
    CleanupStack::PopAndDestroy(crypto);
    }
}
```

14.7 Summary

This chapter covered the important concepts behind the ECOM architecture, which provides services to locate, resolve, instantiate and manage instances of polymorphic plug-in DLLs.

The chapter discussed how to define an abstract interface and how to implement it in an ECOM plug-in DLL. The example showed how to implement two concrete implementation classes for a fictional cryptographic interface, within a single polymorphic plug-in DLL. For this reason, an ECOM plug-in DLL is also known as an "ECOM Interface Implementation Collection", and assigned a `targettype` of `ECOMIIC`.

The chapter also described how an interface client can use factory methods provided by the interface to instantiate concrete instances thereof, without having any knowledge of ECOM itself. The factory instantiation functions, provided by the interface, call the ECOM framework using `REComSession::CreateImplementationL()`. The ECOM framework uses a default or customized resolver to examine the registry of interface implementations and instantiate an object of the appropriate class. An interface client should simply include the header files which define the interface and link against `ECOM.lib`.

Chapter 13 discusses generic polymorphic plug-in DLLs and Symbian OS DLL loading in more detail.

15

Panics

We experience moments absolutely free from worry. These brief respites are called panic

Cullen Hightower

One dictionary definition of panic is "a sudden overpowering fright; sudden unreasoning terror often accompanied by mass flight". The word panic itself derives from Pan, the Greek god of nature and goatherds, who was half man and half goat. He was considered by some to represent an image of the devil. Panic was said to resemble the mental or emotional state induced by him.

On Symbian OS, when a thread is panicked, the code in it stops running. Panics are used to highlight a programming error in the most noticeable way, stopping the thread to ensure that the code is fixed, rather than potentially causing serious problems by continuing to run. There is no recovery from a panic. Unlike a leave, a panic can't be trapped. A panic is terminal.

If a panic occurs in the main thread of a process, the entire process in which the thread runs will terminate. If a panic occurs in a secondary thread, it is only that thread which closes. However, on hardware, if any thread is deemed to be a system thread, i.e. essential for the system to run, a panic in the thread will reboot the phone.[1] Otherwise, if it's not a system thread, on target hardware and in release builds on the Windows emulator, the end result of a panic is seen as a "Program closed" message box, which displays the process name, along with a panic category and error code. In debug emulator builds, you can choose to break into the code to debug the cause of the panic – this is known as just-in-time debugging.

---

[1] Forcing a reboot if a system-critical thread panics is a policy decision rather than a shortcoming of the operating system.

Panics are used to highlight programming errors by terminating the thread (and, if it is the main thread, the process in which it runs).

15.1 Just-In-Time Debugging

On the Windows emulator, in debug builds, you can use the `User::SetJustInTime()` system function to choose whether a panic kills just the thread or the whole emulator. By default, just-in-time debugging is enabled and a panic will terminate the entire program and enter the debugger. You can disable this by calling `User::SetJustInTime(EFalse)`, whereupon the panic will appear as it does in release builds and simply terminate the thread in which it occurred with an appropriate message box. Just-in-time debugging can be re-enabled by calling `User::SetJustInTime(ETrue)`.

When just-in-time debugging is enabled, the panic calls a function called `ThreadPanicBreakPoint()`. This function presents the debugger with an information structure which contains information about the ID and name of the thread that panicked, the panic category and reason (which are described shortly), and the name of the thread that caused the panic. The function then breaks into the code and launches the debugger within the context of the function that called the panic, using `__asm int 3`. You can use the debugger to look through the call stack to see where the panic arose and examine the appropriate state.

There are some subtle differences in behavior between Symbian OS v8.0, which contains the new hard real-time kernel (known at the time of going to press as "EKA2", which stands for "EPOC Kernel Architecture 2") and previous releases of Symbian OS (EKA1).

Panics on EKA1

A call to the static function `User::Panic()` panics the currently running thread. A thread may panic any other thread in the system by acquiring an `RThread` handle to it (as discussed in Chapter 10) and calling `RThread::Panic()`. Both functions take two parameters: a panic category string, which is limited to 16 characters, and an error code, expressed as a `TInt`.

Panics on EKA2

A call to the static function `User::Panic()` panics the currently running thread. A thread may panic any thread in the *same* process by calling `RThread::Panic()`, but can no longer panic threads in any *other* process.[2] The panic functions take a panic category string, which can contain an unlimited number of characters, and an error code, expressed, as usual, as a `TInt`.

15.2 Good Panic Style

Even without breaking into the debugger, you should still be able to track down the cause of a panic using the panic category string and error number. It's good style to make your panic category string descriptive and unique, so other developers can locate the string in your header files, and with it, any associated panic error codes (which should also have suitably descriptive names).

Thus, you might have a general panic header file for your library which includes the following:

```
// ClangerPanic.h
#ifndef __CLANGERPANIC_H__
#define __CLANGERPANIC_H__

#include <e32base.h>

_LIT(KClangerPanic, "CLANGER-ENGINE");

enum TClangerEnginePanic
    {
    ECorruptBlueStringPudding, // =0,
    EIronChickenNotInitialized,// =1,
    EInvalidClangerSetting     // =2
    };

static void Panic(TClangerEnginePanic aCategory);
#endif // __CLANGERPANIC_H__
```

Which defines the `Panic()` function separately as follows:

```
static void Panic(TClangerEnginePanic aCategory)
    {
    User::Panic(KClangerPanic, aCategory);
    }
```

(You'll notice by the way that the panic enumeration is a T Class because an enumeration is a type. Chapter 1 discusses the differences between class types on Symbian OS and how to use them.)

When the library code is passed invalid arguments, it may invoke `Panic()` with the appropriate error code, resulting in a panic and

[2] Except where a server thread uses `RMessagePtr` to panic a misbehaving client thread.

termination of the thread in which it is running. The category and error will be reported, and may then be traced back by searching the library's header files for "CLANGER-ENGINE", located inside ClangerPanic.h. You'll see I've commented each error's enum value with its associated number, just to make the lookup easier. I've tried to give each a descriptive name, though obviously they could be further documented, using in-source comments, for clarity.

Of course, if a client programmer has access to the source code for clanger.dll, they can also search it for calls to Panic() which use a particular error value, to track down where a panic originated.

15.3 Symbian OS Panic Categories

Symbian OS itself has a series of well-documented panic categories and associated error values. You can find details of platform-specific panics in your preferred SDK. From Symbian OS v7.0, there is a special Panics section in the C++ API Reference of each SDK, which contains a comprehensive list of the Symbian OS system panics. Typical values you may encounter include:

- KERN-EXEC 3 – raised by an unhandled exception (such as an access violation caused, for example, by dereferencing NULL, memory misalignment or execution of an invalid instruction) inside a system call to the kernel executive; if an unhandled exception occurs inside code which is instead executing in user mode, the panic is seen as USER-EXEC 3
- E32USER-CBASE 46 – raised by the active scheduler as a result of a stray signal (described further in Chapters 8 and 9)
- E32USER-CBASE 90 – raised by the cleanup stack when the object specified to be popped off is not the next object on the stack. The following code illustrates the cause of the panic; this issue is described more fully in Chapter 3, which describes the cleanup stack.

```
class CExample; // defined elsewhere

void CauseAPanicL()
    {
    TInt val = 1;
    CExample* ptr1 = new (ELeave) CExample();
    CleanupStack::PushL(ptr1);
    CExample* ptr2 = new (ELeave) CExample();
    CleanupStack::PushL(ptr2);
    ...
    CleanupStack::Pop(ptr1); // Panics with E32USER-CBASE 90 here...
```

```
CleanupStack::Pop(ptr2); // ...so the code never gets here
}
```

As a user, if making the chess move you've carefully worked out causes your game to panic and close, you're probably not interested in the category or error code. Such precise details are irrelevant and the "program closed" dialog is more irritating than helpful.

Don't use panics except as a means to eliminate programming errors, for example by using them in assertion statements. Panicking cannot be seen as useful functionality for properly debugged software; a panic is more likely to annoy users than inform them.

15.4 Panicking Another Thread

I've shown how to call `User::Panic()` when you want the thread to panic itself. But how, and when, should you use `RThread::Panic()`? As I described earlier, the `RThread::Panic()` function can be used to kill another thread and indicate that the thread had a programming error. Typically this is used by a server to panic a client thread when the client passes a badly-formed request. You can think of it as server self-defense; rather than go ahead and attempt to read or write to a bad descriptor, the server handles the client's programming error gracefully by panicking the client thread. It is left in a good state which it would not have been if it had just attempted to use the bad descriptor. Generally, in a case like this, the malformed client request has occurred because of a bug in the client code, but this strategy also protects against more malicious "denial of service" attacks in which a client may deliberately pass a badly-formed or unrecognized request to a server to try to crash it. The Symbian OS client–server architecture is discussed in Chapters 11 and 12; the latter chapter includes sample code that illustrates how a server can panic its client.

If you have a handle to a thread which has panicked, you can determine the exit reason and category using `ExitReason()` and `Exit-Category()`. This can be useful if you want to write test code to check that the appropriate panics occur, say from assertions in your library code, to defend it against invalid parameters. Since you can't "catch" a panic, it's not as straightforward as running the code in the main thread, passing in the parameters and checking that the code panics with the correct value. The checking code would never run, because the panic would terminate the main thread.

A solution is to run deliberately badly-behaved client code in a separate test thread, programmatically checking the resulting exit reasons and categories of the panicked thread against those you would expect to have occurred. You should disable just-in-time debugging for the duration of the test, so that only the test thread, rather than the emulator, is terminated. For example:

```
enum TChilliStrength
    {
    ESweetPepper,
    EJalapeno,
    EScotchBonnet
    };

void EatChilli(TChilliStrength aStrength)
    {
    _LIT(KTooStrong, "Too Strong!");
    __ASSERT_ALWAYS(EScotchBonnet!=aStrength,
            User::Panic(KTooStrong, KErrAbort);
    ... // Omitted for clarity
    }

// Thread function
TInt TestPanics(TAny* /*aData*/)
    {// A panic occurs if code is called incorrectly
    EatChilli(EScotchBonnet);
    return (KErrNone);
    }

void TestDefence()
    {
    // Save current just-in-time status
    TBool jitEnabled = User::JustInTime();
    // Disable just-in-time debugging for this test
    User::SetJustInTime(EFalse);

    _LIT(KPanicThread, "PanicThread");
    // Create a separate thread in which to run the panic testing
    RThread testThread;
    TInt r = testThread.Create(KPanicThread, TestPanics,
            KDefaultStackSize, NULL, NULL);

    ASSERT(KErrNone==r);
    // Request notification of testThread's death (see Chapter 10)
    TRequestStatus tStatus;
    testThread.Logon(tStatus);
    testThread.Resume();
    User::WaitForRequest(tStatus); // Wait until the thread dies

    ASSERT(testThread.ExitType()==EExitPanic);
    // Test the panic reason is as expected
    ASSERT(testThread.ExitReason()==KErrAbort);

    testThread.Close();
    // Set just-in-time back to previous setting
    User::SetJustInTime(jitEnabled);
    }
```

15.5 Faults, Leaves and Panics

A fault is raised if a critical error occurs such that the operating system cannot continue normal operation. On hardware, this results in a reboot. A fault can only occur in kernel-side code or a thread which is essential to the system, for example the file server, so typically you will not encounter them unless you are writing device drivers or uncover a bug in the OS. In effect, a fault is another name for a serious system panic.

Chapter 2 discusses leaves in more detail, but, in essence, they occur under exceptional conditions such as out of memory or the absence of a communications link. Unlike a panic, a leave should always be caught ("trapped") by code somewhere in the call stack, because there should always be a top-level `TRAP`. However, if a leave is not caught, this implies that the top-level `TRAP` is absent (a programming error) and the thread will be panicked and terminate.

15.6 Summary

This chapter discussed the use of panics to terminate the flow of execution of a thread. Panics cannot be "caught" like an exception and are severe, resulting in a poor user experience. For that reason, panics are only useful to track down programming errors and, on Symbian OS, are typically combined with an assertion statement, as discussed in the next chapter.

This chapter described the best way to identify panics and illustrated how to test the panics that you've added to your own code for defensive programming. It gave a few examples of commonly-encountered system panics and directed you to the Panics section of the system documentation for a detailed listing of Symbian OS system panics.

16

Bug Detection Using Assertions

On the contrary!

As Ibsen lay dying, his nurse brought him some visitors. "Our patient is feeling much better today," she told them. Ibsen woke up, made the exclamation above, and died.

In C++, assertions are used to check that assumptions made about code are correct and that the state, for example, of objects, function parameters or return values, is as expected. Typically, an assertion evaluates a statement and, if it is false, halts execution of the code and perhaps prints out a message indicating what failed the test, or where in code the failure occurred.

On Symbian OS, you'll find the definition of two assertion macros[1] in `e32def.h`:

```
#define __ASSERT_ALWAYS(c,p) (void)((c)||(p,0))
...
#if defined(_DEBUG)
#define __ASSERT_DEBUG(c,p) (void)((c)||(p,0))
#endif
```

As you can see from the definition, if the assertion of condition `c` is false, procedure `p` is called; this should always halt the flow of execution, typically by panicking (panics are described in detail in Chapter 15). You can apply the assertion either in debug code only or in both debug and release builds. I'll discuss how you decide which is appropriate later in the chapter.

---

[1] At first sight, these definitions seem more complex than they need to be, when the following simpler definition could be used:

```
#define __ASSERT_ALWAYS(c,p) ((c)||(p))
```

The reason for the `(p,0)` expression is for cases where the type returned from `p` is `void` (the case when `p` is a typical `Panic()` function) or a value that can't be converted to an integer type for evaluation. The cast to `void` is present to prevent the return value of the expression being used inadvertently.

You'll notice that the assertion macros do not panic by default, but allow you to specify what procedure to call should the assertion fail. This gives you more control, but you should always terminate the running code and flag up the failure, rather than return an error or leave. Assertions help you detect invalid states or bad program logic so you can fix your code as early as possible. It makes sense to stop the code at the point of error, thus forcing you to fix the problem (or remove the assertion statement if your assumption is invalid). If the assertion simply returns an error on failure, not only does it alter the program flow, but it also makes it harder to track down the bug.

You should always raise a panic when an assertion statement fails.

16.1 __ASSERT_DEBUG

Here's one example of how to use the debug assertion macro:

```
void CTestClass::TestValue(TInt aValue)
    {
    #ifdef _DEBUG
    _LIT(KPanicDescriptor, "TestValue"); // Literal descriptor
    #endif
    __ASSERT_DEBUG((aValue>=0), User::Panic(KMyPanicDescriptor,
        KErrArgument));
    ...
    }
```

Of course, this is somewhat awkward, especially if you expect to use a number of assertions to validate your code, so you'll probably define a panic utility function for your module, with its own panic category string and a set of panic enumerators specific to the class. So, for example, you'd add the following enumeration to `CTestClass`, so as not to pollute the global namespace:

```
enum TTestClassPanic
    {
    EInvalidData,        // =0
    EUninitializedValue // =1
    ...
    };
```

Then define a panic function, either as a member of the class or as a static function within the file containing the implementation of the class:

```
static void Panic(TInt aCategory)
    {
    _LIT(KTestClassPanic, "CTestClass-Panic");
    User::Panic(KTestClassPanic, aCategory);
    }
```

You could then write the assertion in `TestValue()` as follows:

```
void CTestClass::TestValue(TInt aValue)
    {
    __ASSERT_DEBUG((aValue> =0), Panic(EInvalidTestValueInput));
    ...
    }
```

The advantage of using an identifiable panic descriptor and enumerated values for different assertion conditions is traceability, both for yourself and clients of your code, when an assertion fails and a panic occurs. This is particularly useful for others using your libraries, since they may not have access to your code in its entirety, but merely to the header files. If your panic string is clear and unique, they should be able to locate the appropriate class and use the panic category enumeration to find the associated failure, which you will have named and documented clearly to explain why the assertion failed.

There may be cases where there's nothing more a client programmer can do other than report the bug to you, the author of the code; alternatively, the problem could be down to their misuse of your library, which they'll be able to correct. I'll discuss the pros and cons of using assertions to protect your code against badly-programmed calling code later in this chapter.

If you don't want or need an extensive set of enumerated panic values, and you don't expect external callers to need to trace a panic, you may consider using a more lightweight and anonymous assertion. A good example of this is to test the internal state of an object, which could not possibly be modified by an external caller, and thus should always be valid unless you have a bug in your code. Assertions can be added early in the development process, but left in the code, in debug builds, to validate the code as it is maintained and refactored. In these cases, you may consider using the `ASSERT` macro, defined in `e32def.h` as follows:

```
#define ASSERT(x) __ASSERT_DEBUG(x, User::Invariant())
```

I like this macro because it doesn't need you to provide a panic category or descriptor. If condition `x` is false, in debug builds only, it calls `User::Invariant()` which itself panics with category `USER` and reason 0. The macro can be used as follows:

```
ASSERT(iClanger>0);
```

As an alternative to using `ASSERT` to test the internal state of your object, you may wish to consider using the `__TEST_INVARIANT` macro, which I discuss in more detail in Chapter 17.

An alternative, useful definition of `ASSERT` which you may see in some Symbian OS code is as follows:

```
#ifdef _DEBUG
#ifdef ASSERT
#undef ASSERT
#endif
#define __ASSERT_FILE__(s) _LIT(KPanicFileName,s)
#define __ASSERT_PANIC__(l) User::Panic(KPanicFileName().Right(12),l)
#define ASSERT(x) { __ASSERT_FILE__(__FILE__);
      __ASSERT_DEBUG(x, __ASSERT_PANIC__(__LINE__) ); }
#endif
```

This slightly alarming construction is actually quite simple; in debug builds, if condition `x` is false, the code is halted by a panic identifying the exact place in code (in the panic descriptor – which contains the last 12 characters of the filename) and the panic category (which contains the line of code at which the assertion failed). The disadvantage of using this construct is that you are coupling the compiled binary directly to the source file. You cannot later modify your code file, even to make non-functional changes to comments or white space lines, without recompiling it to update the assertion statements. The resulting binary will differ from the original, regardless of the nature of the changes. Depending on how you deliver your code, this limitation may prohibit you from using this macro.

Let's move on from how to use the Symbian OS assertion syntax to consider when you should use assertions and, perhaps more importantly, when you should not.

Firstly, don't put code with side effects into assertion statements. By this, I mean code which is evaluated before a condition can be verified.

For example:

```
__ASSERT_DEBUG(FunctionReturningTrue(), Panic(EUnexpectedReturnValue));
__ASSERT_DEBUG(++index<=KMaxValue, Panic(EInvalidIndex));
```

The reason for this is clear; the code may well behave as you expect in debug mode, but in release builds the assertion statements are removed by the preprocessor, and with them potentially vital steps in your programming logic. Rather than use the abbreviated cases above, you should perform the evaluations first and then pass the returned values into the assertion. You should follow this rule for both `__ASSERT_DEBUG` and `__ASSERT_ALWAYS` statements, despite the fact that the latter are compiled into release code, because, while you may initially decide the assertion applies in release builds, this may change during the development or maintenance process. You could be storing up a future bug for the sake of avoiding an extra line of code.

You must also make a clear distinction between programming errors ("bugs") and exceptional conditions. Examples of bugs might be contradictory assumptions, unexpected design errors or genuine implementation errors, such as writing off the end of an array or trying to write to a file before opening it. These are persistent, unrecoverable errors which should be detected and corrected in your code at the earliest opportunity. An exceptional condition is different in that it may legitimately arise, although it is rare (hence the term "exceptional") and is not consistent with typical or expected execution. It is not possible to stop exceptions occurring, so your code should implement a graceful recovery strategy. A good example of an exceptional condition that may occur on Symbian OS is an out-of-memory failure, because it is designed to run constantly on devices with limited resources for long periods of time without a system reset.

To distinguish between bugs and exceptions, you should consider the following question. Can a scenario arise legitimately, and if it can, is there anything you should or could do to handle it? If your answer is "yes", you're looking at an exceptional condition – on Symbian OS, this is exhibited as a leave (leaving is discussed in Chapter 2). If the answer is "no", you should consider the situation to be caused by a bug which should be tracked down and fixed. The rest of this chapter will focus on the use of assertions to highlight such programming errors.

When code encounters a bug, it should be flagged up at the point at which it occurs so it can be fixed, rather than handled or ignored (which can at best complicate the issue and, at worst, make the bug more difficult to find or introduce additional defects as you "code around it"). You could consider assertions as an annoying colleague, leaning over your shoulder pointing out defects for you as your code runs. They don't prevent problems, but make them obvious as they arise so you can fix them.

If you add assertion statements liberally as you write code, they effectively document assumptions you make about your program logic and may, in addition, flag up unexpected problems. As you consider which assertions to apply, you are actually asking yourself what implicit assumptions apply to the code and how you can test them. By thinking about each piece of code you write in this way, you may well discover other conditions to test or eliminate that would not have been immediately obvious. Frequent application of assertions as you code can thus help you to pre-empt bugs, as well as catch those already in existence.

And there's more! Another benefit of assertions is the confidence that your code is behaving correctly, and that you are not ignoring defects which may later manifest themselves where they are hard to track down, for example intermittently or through behavior seemingly unrelated to the code that contains the error. Say you write some library code containing no assertions and then create some code to test it, which runs and returns no errors; are you confident that everything behaves as you expect and the test code checks every boundary condition? Are you sure? Certain?

Adding assertions to test fundamental assumptions in your code will show you immediately if it is swallowing or masking defects – and it should give you more confidence that the test results are valid. Sure, you might hit some unpleasant surprises as the test code runs for the first time, but once you've ironed out any failures, you can be more confident about its overall quality. What's more, the addition of assertions also protects your code against any regressions that may be introduced during maintenance and refactoring but which would not otherwise be picked up by your test code. What more could you ask?

The cases so far could be considered as "self defense", in that I've discussed using assertions to catch bugs in *your* code. Let's move on to consider defensive programming in general. Defensive programming is not about retorting "It works OK on my machine" after being informed that your code doesn't work as expected. It's based on defending your code against irresponsible use or downright abuse by code that calls it. Defensive code protects functions against invalid input, by inspecting data passed in and rejecting corrupt or otherwise flawed parameters, such as strings that are too long or out-of-range numerical values.

You'll need to consider how to handle bad parameters depending on how your code is called; for example, you may want to assert that the data is good, terminating with a panic if it is not. Alternatively, you may decide to continue the flow of execution, so instead of assertions, you'll check each incoming parameter (e.g. using `if` statements) and return to the caller if invalid data is detected – either with an error value or a leave code. Another method would be to check incoming data and, if a parameter is invalid, substitute it with a default parameter or continue with the closest legal value. What you don't want to do is ignore invalid input and carry on regardless, since this could lead to problems later on, such as data corruption. Whatever method you use to handle illegal input, it should be consistent throughout your code.

Your clients should be testing with debug versions of your libraries and thus you could use `__ASSERT_DEBUG` statements to alert them of invalid input, or other misuse, so they can correct it.

__ASSERT_DEBUG assertions can be added early in the development process to highlight programming errors and can be left in to validate the code as it is maintained and refactored – acting as a means to "design by contract".

16.2 __ASSERT_ALWAYS

You still need to be defensive by checking for illegal usage in release builds too, but the case for using `__ASSERT_ALWAYS` isn't clear cut. Remember, your assertions will terminate the flow of execution and panic the library, displaying a nasty "program closed" dialog to the user, which is generally best avoided where possible. Additionally, you should consider the impact on the speed and size of your code if you apply assertion statements liberally in release builds.

If you decide not to use `__ASSERT_ALWAYS` to check incoming values, you should use another defensive technique to guard against illegal input, such as a set of `if` statements to check values and return error codes or leave when data is unusable. You could use these in combination with a set of `__ASSERT_DEBUG` statements to alert the client programmer to invalid use in debug builds, but often it is preferable to keep the flow of execution the same in both debug and release builds. In such cases, I suggest you don't use debug assertions to check input, but instead use `if` statement checking in both modes, and document each expected return value for your functions. Client programmers should understand their responsibility to interpret the return value and act accordingly. I'll illustrate this with an example later in this chapter.

To determine whether you should use `__ASSERT_ALWAYS` or another, less terminal, defense, I recommend that you consider whether the calling code may be able to take a different action if you do return an error. Invalid input is a bug from the perspective of your code, but may be caused by an exceptional condition in the calling code which can be handled.

A simplistic example would be a call to your code to open and write to a file, where the caller passes in the full file name and path, as well as the data to be written to the file. If the file does not exist, it is probably more appropriate to return this information to the caller through a returned error code or leave value than to assert in a release build. Client code can then anticipate this and deal with it, without the need for your library to panic and alarm the user accordingly.

Here's an example of this scenario which illustrates how to defend against illegal parameters without assertions. This code returns an error to allow the caller to recover if they pass in an invalid parameter. Of course, if the calling code is written in such a way that each parameter should be correct and that only a bug could result in them being invalid, it can assert that the return value from a call to `WriteToFile()` is `KErrNone`. On the other hand, if it's an exceptional circumstance that the file is missing or the data is non-existent, it can handle it gracefully.

```
TInt CTestClass::WriteToFile(const TDesC& aFilename,
      const TDesC8& aData)
   {
   TInt r = KErrNone;
   if (KNullDesC8==aData)
      {// No data to write - invalid!
      r = KErrArgument;
      }
   else
      {
      RFile file;
      __ASSERT_DEBUG(iFs, Panic(EUninitializedValue));
      r = file.Open(iFs, aFilename, EFileWrite);
      if (KErrNone==r)
         {   // Only executes if the file can be opened
         ... // Writes aData to file, closes file etc
         }
      }

   return (r);
   }
```

You'll notice that I've included an `__ASSERT_DEBUG` statement to verify internal state in my code and catch any defects (such as attempting to use the file server handle before it has been initialized) in the test phase.

One case where you may prefer to use `__ASSERT_ALWAYS` to protect your code against illegal input is where that input could only have arisen through a bug in calling code and will cause "bad things", such as memory corruption, to occur. You could return an error to the caller, but it's probably clearer for the calling code if you flag up the problem so it can be fixed. A good example of this is in the Symbian OS array classes (`RArray` and `RPointerArray`), which have `__ASSERT_ALWAYS` guards to prevent a caller passing an invalid index to methods that access the array. The class provides functions to determine the size of the array, so if a caller attempts to write off the end of the array, it can only be doing so because of a bug.

Likewise, in the code above, if the context of the function means that the second parameter, `aData`, should never be an empty string, you can replace the first `if` statement check with an `__ASSERT_ALWAYS` statement. But this assumes knowledge of how clients expect to call it and reduces the option for reuse at a later stage, should this condition no longer apply.

If the caller is passing data from a source that it does not directly control, say a communications link, there is always a possibility for invalid input to your code. In these circumstances, it's better to handle bad incoming data by returning an error or leaving. It is unusual to use assertions under these conditions, although code may occasionally need to do so, depending on the circumstances in which it is used. Whatever the decision, ignoring the problem of illegal input is not an option!

Use defensive coding techniques to protect your functions against invalid input from calling code. `__ASSERT_ALWAYS` should be used to protect against illegal input that can only ever have arisen through a bug in the caller.

16.3 Summary

This chapter discussed assertion statements on Symbian OS in terms of how they work, what they do and how you should use them. Their primary purpose is for you to verify program logic as you write code and to detect bugs at the point they occur so you can find and fix them. For this reason, assertion statements must terminate the flow of execution upon failure, typically with a panic.

The chapter recommended using `__ASSERT_DEBUG` or `ASSERT` statements liberally in your code to check its internal state. You should expect to test your code thoroughly and fix defects before release, so you shouldn't need to use `__ASSERT_ALWAYS` to check the internals of your code. A user, or another software developer using your code, expects you to have debugged your code; they don't want to do it for you. Furthermore, you should think carefully before adding assertion statements to release code because of the added code size and extra execution overhead associated with them.

You should make the distinction between bugs and exceptions (failures occurring under exceptional circumstances). While you can use assertions to catch bugs, you should handle exceptions gracefully, since they may occur legitimately in your code, albeit as an atypical path of execution. This point extends to your client's code – you should program defensively and consider whether invalid input has arisen because of a bug or exception in the caller. If it can only be due to a bug, it is acceptable to use `__ASSERT_ALWAYS` statements in your code to indicate to the client that they have a programming error which needs fixing. However, since release build assertion statements have a cost in terms of size, speed and ugly code termination, I advise you to consider carefully before using them against invalid input data which may have arisen from an exception that the caller can handle more gracefully.

This chapter illustrated some aspects of defensive programming besides the use of assertion statements, but added this note of caution. You should consider carefully how much defensive code to use, and whether it varies between debug and release builds. It can create additional complexity in your code, leaving it open to its own set of bugs, and, if you check parameter data for every possible error, it can also make your code slow and bloated. You should take care to use defensive techniques where they are most effective, and document them clearly so the client can build their own bug catching and exception handling around them.

The paradox is that you want problems to be noticeable so they are flagged up during development, but you want them to be inconspicuous in your production code. You should consider your approach to defensive code and assertions appropriately for each project you work on. In all cases, it's best to keep it consistent and consider error handling as an important issue to be determined and defined at an architectural level.

The next chapter discusses useful debug macros and test classes on Symbian OS for tracking down programming errors such as memory leaks and invalid internal state. You can find more information about handling leaves (Symbian OS exceptions) in Chapter 2.

17

Debug Macros and Test Classes

If you have built castles in the air, your work need not be lost; that is where they should be. Now put foundations under them

Henry David Thoreau (Walden)

Memory management is an important issue on Symbian OS, which has a number of mechanisms to assist you in writing memory leak-proof code. This chapter starts by discussing a set of debug macros that allow you to check that your code is managing heap memory correctly (you will find them documented in your SDK under "Memory Allocation"). It will also discuss other useful debug macros and the `RTest` and `RDebug` classes.

17.1 Heap-Checking Macros

There are macros which can be used to check that you are not leaking any memory under normal conditions and others that can be used to simulate out-of-memory (OOM) conditions. It is advisable to check your code using each macro type, to ensure both that it is leave-safe and that it can handle OOM gracefully (that is, to ensure that heap memory is not orphaned by a leave occurring under OOM as I discussed in Chapter 2). This kind of testing is known as "Alloc Heaven Testing".

Here are the macros, as defined in `e32def.h`. They are only compiled into debug builds, and are ignored by release builds, so they can be left in your production code without having any impact on the code size or speed.

```
#define __UHEAP_MARK User::__DbgMarkStart(RHeap::EUser)
#define __UHEAP_MARKEND User::__DbgMarkEnd(RHeap::EUser,0)
#define __UHEAP_MARKENDC(aCount)
        User::__DbgMarkEnd(RHeap::EUser,aCount)
#define __UHEAP_FAILNEXT(aCount)
        User::__DbgSetAllocFail(RHeap::EUser,RHeap::EFailNext,aCount)
#define __UHEAP_SETFAIL(aType,aValue)
        User::__DbgSetAllocFail(RHeap::EUser,aType,aValue)
#define __UHEAP_RESET
        User::__DbgSetAllocFail(RHeap::EUser,RHeap::ENone,1)
```

The __UHEAP_MARK and __UHEAP_MARKEND macros verify that the default user heap is consistent. The check is started by using __UHEAP_MARK and a following call to __UHEAP_MARKEND performs the verification. If any heap cells have been allocated since __UHEAP_MARK and not freed back to the heap, __UHEAP_MARKEND will cause a panic.[1] These macros can be nested inside each other and used anywhere in your code. You must take care to match the pairs, because a __UHEAP_MARKEND without a prior __UHEAP_MARK causes a panic.

__UHEAP_FAILNEXT simulates a heap failure (i.e. out of memory). The macro takes a TInt parameter which specifies which memory allocation call will fail (if the parameter is 1, the first heap allocation fails, if it's 2, the second allocation fails, and so on). The __UHEAP_RESET macro cancels the failure checking.

Consider the following example, where a method CTest::SomethingL() makes several memory allocations internally. To test the leave safety of the method, it is necessary to check that failure of each of these allocations is handled correctly. The best way to do this is to call SomethingL() inside a loop as follows:

1. Set the __UHEAP_FAILNEXT parameter.
2. Use __UHEAP_MARK to mark the heap.
3. Call CTest::SomethingL() inside a TRAP macro.
4. Call __UHEAP_MARKEND to check that no memory has been leaked.
5. Check the error from the TRAP; if it's KErrNoMemory, you should go round the loop again, incrementing the number of allocations which succeed before the failure point. If it's KErrNone, all the allocation points inside SomethingL() have been tested and the test is complete.

```
void TestLeaveSafetyL()
   {
   // extra check to demonstrate that these macros can be nested
   __UHEAP_MARK;

   CTest* test = CTest::NewLC();
   for (TInt index=1;;index++)
      {
      __UHEAP_FAILNEXT(index);
      __UHEAP_MARK;

      // This is the function we are OOM testing
```

[1] The call to User::__DbgMarkEnd() will raise a panic if the heap is inconsistent. The panic raised is ALLOC nnnnnnnn, where nnnnnnnn is a hexadecimal pointer to the first orphaned heap cell. I'll discuss how to track down what is causing a panic of this type shortly.

```
    TRAPD(r, test->SomethingL());

    __UHEAP_MARKEND; // Panics if the heap memory is leaked

    // Stop the loop if r! =KErrNoMemory
    if (KErrNoMemory!=r)
        {
        User::LeaveIfError(r);
        break; // For cases where r==KErrNone
        }
    }
    CleanupStack::PopAndDestroy(test);
    __UHEAP_MARKEND;
    __UHEAP_RESET;
}
```

If your code fails to propagate allocation failures, then the loop above will not work, although your code will appear to be leave-safe. Suppose that `CTest::SomethingL()` makes five heap allocations, but when the third allocation fails, the error is "swallowed" and not propagated back to the caller of `SomethingL()` as a leave. The loop will terminate at the third allocation and the test will appear to have been successful. However, if a leave from the fourth allocation causes a memory leak, this will go undetected. It is thus important to make sure that your code always propagates any heap allocation failures.

Another problem is that it is possible that some code will allocate memory which is not freed before the `__UHEAP_MARKEND` macro. A good example is the cleanup stack, which may legitimately be expanded as code runs but does not free the memory allocated to it until the code terminates (as I described in Chapter 3). This problem can also occur when using any dynamic container class that supports automatic buffer expansion to hide memory management (Chapter 7 examines these classes in detail).

To prevent a false memory leak panic in your test code, it is advisable to expand the cleanup stack beyond what it is likely to need before the first `__UHEAP_MARK` macro:

```
void PushLotsL()
    {// Expand the cleanup stack for 500 pointers
        {
        TInt* dummy = NULL;
        for (TInt index =0; index<500; index++)
        CleanupStack::PushL(dummy);
        }
    CleanupStack::Pop(500);
    }
```

The `__UHEAP_X` (where `X` = `MARK`, `MARKEND`, `MARKENDC`, `FAIL-NEXT`, `SETFAIL` and `RESET`) macros only work on the default heap for the thread from which they are called. If your test code is in a different process to the code you're testing, which is typical when testing client–server code, the macros I've discussed won't work. There are at least three possible options here:

- In addition to the __UHEAP_X macros, there is also a set of equivalent __RHEAP_X macros which perform checking on a specific heap, identified by an RHeap parameter.[2] If you take a handle to the server's heap, you can pass it to these functions to check the server's heap from a separate process. You can get a handle to the server's heap by opening an RThread handle on the server's main thread (using RThread::Open()) and then calling Heap() on this handle to acquire an RHeap pointer.
- Add a debug-only test API to your server which calls the appropriate __UHEAP_X macro server-side. You can then define your own macros which call these methods:

```
RMyServerSession::MarkHeap() calls __UHEAP_MARK
RMyServerSession::MarkHeapEnd() calls __UHEAP_MARKEND
RMyServerSession::FailNextAlloc() calls __UHEAP_FAILNEXT
RMyServerSession::ResetHeap() calls __UHEAP_RESET
```

- Compile the server code directly into your test executable rather than running a client–server IPC framework. This is necessary where you don't control the client–server API, but is a cumbersome solution in general.

The Symbian OS header file, e32def.h, also defines the __UHEAP_CHECK, __UHEAP_CHECKALL and __UHEAP_MARKENDC macros. These are useful if you want to verify the number of objects currently allocated on the default heap. __UHEAP_CHECKALL takes a TInt parameter, checks it against the number of current heap allocations and raises a panic if the numbers differ. __UHEAP_CHECK is similar, except it counts only those allocations since the last __UHEAP_MARK call. Finally, __UHEAP_MARKENDC has a similar role to __UHEAP_MARKEND, but rather than checking that nothing is left on the heap, it checks that the number of outstanding heap allocations matches the TInt parameter passed to it.

If you surround code which leaks heap memory with a pair of __UHEAP_MARK and __UHEAP_MARKEND macros, an ALLOC panic will be raised by User::__DbgMarkEnd(), which is called by __UHEAP_MARKEND. So what should you do if you encounter such a panic? How do you debug the code to find out what was leaked so you can fix the problem?

---

[2] Incidentally, Symbian OS also provides a set of __KHEAP_X macros, which look identical to the __UHEAP_X macros, but can be used to check the kernel heap, for example when writing device driver code.

The panic reason for the `ALLOC` panic contains a hexadecimal value which indicates the address of the first orphaned heap cell. First of all, you should try to get more information about the type of the object that was left orphaned on the heap. I'm assuming that you can reproduce the panic using the Windows emulator running in a debugger. With just-in-time debugging enabled, as described in Chapter 15, you should run your test code to the breakpoint just after the panic. You should then be able to find the panic reason, expressed as an 8-digit value (e.g. `c212d698`). If you type this value into a watch window, prefixing it with `0x`, you'll have a view of the orphaned heap cell.

You may be able to recognize the object simply by inspecting the contents of the memory in the debugger. Otherwise, it's worth casting the address in the watch window to (`CBase*`) to see if you can get more information about the object. If it's a C class object, that is, `CBase*`-derived, the cast allows the debugger to determine the class of the object from its virtual function table pointer. Since C classes are always heap-based, there's a fair chance that the orphaned object can be identified in this way. If there's no information to be gleaned from a `CBase*` cast, you can also try casting to `HBufC*` or any of the other classes that you know are used on the heap in the code in question.

Once you know the type of the orphaned object, that's often sufficient to help you track down the leak. However, if there are numerous objects of that type, or you can't identify it (perhaps because the leak is occurring in a library which your code is using, rather than in your own code), you may need to use the debugger to add conditional breakpoints for that particular address in memory. With these in place, when you run the code again, it should break at the heap cell whenever it is allocated, de-allocated or modified. You may find that the breakpoints are hit a few times, as the heap cell is allocated and deallocated, but it is only the last allocation that you're interested in. When you've determined the point at which the memory is allocated for the last time, you can inspect the call stack to see what point you're reached in your code. This should help you track down which object is being allocated, and later orphaned, and thus resolve the leak. You can find more information about how to use the debugger to find a memory leak on the Symbian Developer Network website (on ***www.symbian.com/developer***, navigate to the Knowledgebase and search for "memory leak").

It's good practice to use the `__UHEAP_X` macros to verify that your code does not leak memory. You can put `__UHEAP_MARK` and `__UHEAP_MARKEND` pairs in your test code and, since the macros are not compiled into release builds, you can use them to surround areas where leaks may occur in your production code too.

17.2 Object Invariance Macros

The `__DECLARE_TEST` and `__TEST_INVARIANT` macros allow you to check the state of an object, for example at the beginning and end of a complex function. The `__DECLARE_TEST` macro provides a standard way to add a function to a class that allows object invariance to be tested. This function is named `__DbgTestInvariant()` and you should define it as necessary. When performing the invariance testing, rather than call the function directly, you should use the `__TEST_INVARIANT` macro, as I'll show shortly.

`__DECLARE_TEST` is defined as follows in `e32def.h`:

```
#if defined(__DLL__)
#define __DECLARE_TEST public: IMPORT_C void __DbgTestInvariant()
        const; void __DbgTest(TAny *aPtr) const
#else
#define __DECLARE_TEST public: void __DbgTestInvariant() const;
        void __DbgTest(TAny *aPtr) const
#endif
```

You should add it to any class for which you wish to perform an object invariance test. It's best to add it to the bottom of the class declaration because it uses the `public` access specifier. There are separate definitions for DLLs and EXEs because the function must be exported from a DLL so it may be called from a separate module. The second function defined by `__DECLARE_TEST`, `__DbgTest()`, is designed to allow test code to access non-public members of the class, if such testing is required. Because any test code will implement this function itself, it is not exported.

The `__DECLARE_TEST` macro is defined to be the same for both release and debug builds, because otherwise the export list for debug builds would have an extra function (`__DbgTestInvariant()`) compared to the release version. This could cause binary compatibility problems, by changing the ordinal numbers of other exported functions in the module. (Chapter 18 discusses binary compatibility in more detail.)

The `__TEST_INVARIANT` macro is defined below; as you can see, it doesn't call `__DbgTestInvariant()` in release builds. However, if you don't want the invariance test code to be built into release builds, you should surround it with preprocessor directives as appropriate to build the invariance testing into debug builds only.

```
#if defined(_DEBUG)
#define __TEST_INVARIANT __DbgTestInvariant()
#else
#define __TEST_INVARIANT
#endif
```

The invariance test function, `__DbgTestInvariant()`, typically checks that the object's state is correct at the beginning and end of a function. If the state is invalid, a panic should be raised by calling `User::Invariant()` (which panics with category `USER`, reason `0`). You may recall, from Chapter 16, that this is the panic raised by the `ASSERT` macro. Thus a simple way to implement the invariance testing in `__DbgTestInvariant()` is to perform the invariance checks inside `ASSERT` statements.

Here's an example of a class that represents a living person and uses the `__DECLARE_TEST` invariance test to verify that:

- the person has a gender
- if the name is set, it is a valid string of length no greater than 250 characters
- the given age in years matches the age calculated for the given date of birth and is no greater than 140 years.

```
class CLivingPerson: public CBase
    {
public:
    enum TGender {EMale, EFemale};
public:
    static CLivingPerson* NewL(const TDesC& aName,
            CLivingPerson::TGender aGender);
    ~CLivingPerson();
public:
    ... // Other methods omitted for clarity
    void SetDOBAndAge(const TTime& aDOB, const TInt aAge);
private:
    CLivingPerson(TGender aGender);
    void ConstructL(const TDesC& aName);
private:
    HBufC* iName;
    TGender iGender;
    TInt iAgeInYears;
    TTime iDOB; // Date of Birth
    __DECLARE_TEST; // Object invariance testing
    };

CLivingPerson::CLivingPerson(TGender aGender)
: iGender(aGender) {}

CLivingPerson::~CLivingPerson()
{delete iName;}

CLivingPerson* CLivingPerson::NewL(const TDesC& aName,
        CLivingPerson::TGender aGender)
    {
    CLivingPerson* me = new (ELeave) CLivingPerson(aGender);
    CleanupStack::PushL(me);
    me->ConstructL(aName);
    CleanupStack::Pop(me);
```

```
    return (me);
    }

void CLivingPerson::ConstructL(const TDesC& aName)
    {
    __TEST_INVARIANT;
    iName = aName.AllocL();
    __TEST_INVARIANT;
    }

void CLivingPerson::SetDOBAndAge(const TTime& aDOB, const TInt aAge)
    {// Store the DOB and age and check object invariance
    __TEST_INVARIANT;
    iDOB=aDOB;
    iAgeInYears=aAge;
    __TEST_INVARIANT;
    }

void CLivingPerson::__DbgTestInvariant() const
    {
    #ifdef _DEBUG // Built into debug code only
    if (iName)
        {// Name should be more than 0 but not longer than 250 chars
        TInt len = iName->Length();
        ASSERT(len>0); // A zero length name is invalid
        ASSERT(len<250); // The name should be less than 250 characters
        }

    // Person should male or female
    ASSERT((EMale==iGender)||(EFemale==iGender));

    if (iDOB>(TTime)0)
        {// DOB is set, check that age is correct and reasonable
        TTime now;
        now.HomeTime();

        TTimeIntervalYears years = now.YearsFrom(iDOB);
        TInt ageCalc = years.Int();  // Calculate age in years today
        ASSERT(iAgeInYears==ageCalc);// Compare with the stored age
        ASSERT(iAgeInYears>0); // Stored value shouldn't be 0 or less
        ASSERT(iAgeInYears<140); // A living person is less than 140
        }
    #endif
    }

void TestPersonL()
    {
    _LIT(KSherlockHolmes, "Sherlock Holmes");
    CLivingPerson* holmes = CLivingPerson::NewL(KSherlockHolmes,
            CLivingPerson::EMale);

    // Holmes was (apparently) born on 6th January 1854, so he would
    // have been 150 years old at the time of writing (2004)
    TDateTime oldTime(1854, EJanuary, 5, 0, 1, 0, 0);
    // This will panic. Holmes is older than 140!
    holmes->SetDOBAndAge(TTime(oldTime), 150);
    delete holmes;
    }
```

One last point: if the class derives from a class which also implements the invariance test function, __DbgTestInvariant() should be called for the base class first, then any checking done on the derived object.

17.3 Console Tests Using RTest

Class RTest is not documented, but is a useful test utility class, used by a number of Symbian OS test harnesses. You'll find the class definition in \epoc32\include\e32test.h. You should bear in mind that, because the RTest API is not published, there are no firm guarantees that it will not change in future releases of Symbian OS.

RTest is useful for test code that runs in a simple console. You can use the class to display text to the screen, wait for input, and check a conditional expression, rather like an assertion, raising a panic if the result is false.

The main methods of RTest are Start(), End() and Next(), which number the tests, and operator(), which evaluates the conditional expression. The first call to Start() begins numbering at 001 and the numbering is incremented each time Next() is called. Nested levels of numbering can be achieved by using additional calls to Start(), as shown in the example. Each call to RTest::Start() must be matched by a call to RTest::End().

The following source code can be used as the basis of a simple console-based test project, using targettype exe, as described in Chapter 13. Remember, if you want to use the cleanup stack in console-based test code, or link against a component which does so, you will need to create a cleanup stack at the beginning of the test, as discussed in Chapter 3. Likewise, if you write or link against any code which uses active objects, you will need to create an active scheduler, as I described in Chapters 8 and 9.

```
#include <e32test.h>
GLDEF_D RTest test(_L("CONSTEST"));

TInt Test1()
    {
    test.Next(_L("Test1"));
    return (KErrNone);
    }

TBool Test2()
    {
    test.Next(_L("Test2"));
    return (ETrue);
    }

void RunTests()
    {
```

```
    test.Start(_L("RunTests()"));
    TInt r = Test1();
    test(KErrNone==r);
    test(Test2());
    test.End();
    }

TInt E32Main()
    {
    __UHEAP_MARK; // Heap checking, as described above
    test.Title();
    test.Start(_L("Console Test"));
    TRAPD(r,RunTests());
    test.End();
    test.Close();
    __UHEAP_MARKEND;
    return (KErrNone);
    }
```

You'll notice that I'm using the `_L` form of a literal descriptor in the test code, despite mentioning in Chapter 5 that it is deprecated because it constructs a temporary stack-based `TPtrC` object at runtime. However, it's acceptable to use it for debug and test code which is never released and, in this test example, I think it's clearer than predefining a set of `_LIT` literal descriptors.

The code above gives the following output:

```
RTEST TITLE: CONSTEST X.XX(XXX) // Build version and release number

Epoc/32 X.XX(XXX)               // Build version and release number

RTEST: Level  001
Next test - Console Test

RTEST: Level  001.01
Next test - RunTests()

RTEST: Level  001.02
Next test - Test1

RTEST: Level  001.03
Next test - Test2
RTEST: SUCCESS : CONSTEST test completed O.K.
```

`RTest::operator()` can be used to evaluate boolean expressions; if the expression is false, a `USER 84` panic occurs. For example, if `Test2()` above is modified to return `EFalse`, running the test gives following output:

```
RTEST TITLE: CONSTEST X.XX(XXX) // Build version and release number

Epoc/32 X.XX(XXX)               // Build version and release number
```

```
RTEST: Level  001
Next test - Console Test

RTEST: Level  001.01
Next test - RunTests()

RTEST: Level  001.02
Next test - Test1

RTEST: Level  001.03
Next test - Test2

RTEST: Level  001.03
: FAIL : CONSTEST failed check 1 in Constest.cpp at line number 21

RTEST: Checkpoint-fail
```

You'll notice this #define statement in e32test.h:

```
#define test(x) __test(x,__LINE__,__FILE__)
```

This is useful, because, if you use a variable name of test for the RTest object as in the example, even if you use the simplest overload of operator() which just takes the conditional expression, the actual overload invoked includes the source file and line number at which the test occurs. However, if you prefer not to display and log this additional information, you should simply give the RTest object a variable name other than test.

All the RTest functions which display data on the screen also send it to the debug output using a call to RDebug::Print(). If you're running the code on the Windows emulator, this function passes the descriptor to the debugger for output. On hardware, the function uses a serial port which can be connected to a PC running a terminal emulator to view the output (for this output, of course, you need a spare serial port on the phone and a connection to a PC). You can set the port that should be used by calling Hal::Set(EDebugPort,aPortNumber). You'll find a debug log can be extremely useful when running your code on the phone as you develop it. Class RDebug is defined in e32svr.h and provides a number of other useful functions, such as support for profiling.

On the emulator, the debug output is also written to file, so you can run a batch of tests and inspect the output of each later. The emulator uses the TEMP environment variable to determine where to store the file, which it names epocwind.out (%TEMP%\epocwind.out).[3]

---

[3] On Windows NT this is usually C:\TEMP, while on versions of Windows later than Windows 2000, each user has a temporary directory under their Local Settings folder (C:\Documents and Settings\UserName\Local Settings\Temp). Of course, you can change the TEMP environment variable to any location you want to make access to this file convenient.

I've not illustrated the use of `RTest::Printf()` and `RTest::Getch()` in the example code, but they can be used in much the way you would expect, to print to the display/debug output and wait for an input character respectively.

Class `RTest` is useful for test code that runs in a simple console, and can be used to check conditional expressions, raising a panic if a result is false. If you want to use the cleanup stack in console-based test code, or link against a component which does so, you will need to create a cleanup stack as discussed in Chapter 3. If you write or link against any code which uses active objects, you will need to create an active scheduler, as described in Chapters 8 and 9.

17.4 Summary

Memory is a limited resource on Symbian OS and must be managed carefully to ensure it is not wasted by memory leaks. In addition, an application must handle gracefully any exceptional conditions arising when memory resources are exhausted. Symbian OS provides a set of debug-only macros that can be added directly to code to check both that it is well-behaved and that it copes with out-of-memory conditions. This chapter described how to use the macros, and how to debug your code to find the root cause when the heap-checking code panics to indicate a heap inconsistency.

The chapter also discussed how to use the object invariance macro that Symbian OS provides to verify the state of an object. The macro can be added to any class and offers the opportunity to write customized debug state-checking for objects of the class.

Finally, the chapter described the `RTest` class, which is useful for console-based test code. The API for `RTest` can be found in `e32test.h` and is not published, so it may be changed without notice but, since it is used internally by Symbian OS test code, it is unlikely to be removed. `RTest` can be used to run test code in a console, and display the results to the screen, debug output and log file. The class also provides a means of testing return values or statements, which behaves rather like an assertion and panics when the check fails.

18

Compatibility

It is only the wisest and the very stupidest who cannot change

Confucius

I've devoted this chapter to compatibility, because it is an important aspect of software engineering. If you deliver your software as a set of interdependent components, each of which evolves independently, you will doubtless encounter compatibility issues at some stage.

For Symbian, most compatibility issues apply to dependent code, such as that belonging to third-party developers, which should ideally work equally well on products available today, and with future versions of Symbian OS. Compatibility must be managed carefully between major Symbian OS product releases. For developers delivering Symbian OS code that is not built into ROM, there is the additional complexity of managing compatibility between releases that are built on the same Symbian OS platform.

Let's consider the basics of delivering a software library – a binary component which performs a well-defined task, such as a set of string manipulation functions or a math library. The software's author typically publishes the component's application programming interface (API) in the form of C++ header files (`.h`), which are used by a customer's compiler, and an import library (`.lib`), which is used by a linker. Often these are also accompanied by written documentation that describes the correct way to use the library. The API presents, in effect, a contract that defines the behavior of the component to those that use it (its dependents).

Over time, the library may be expanded to deliver additional functionality, or it may be enhanced to remove software defects, make it faster or smaller, or conform to a new standard. These modifications may require changes to the API; in effect, the contract between the component and its dependents may need to be re-written. However, before doing this, the impact of those changes on any dependent code must be assessed. The nature and implications of a change must be well-understood because, when a library has a number of dependent components in a system, even an apparently simple modification can have a significant effect.

This is a fundamental issue of software engineering, and a number of techniques for writing extensible but compatible software have been developed. Many of the issues associated with compatibility lie outside the scope of this book, but an understanding of the basics will allow you to appreciate the factors which have shaped the evolution of Symbian OS. This chapter defines some of the basic terminology of compatibility. It then moves on to discuss some of the practical aspects, which will help you to develop code on Symbian OS which can be extended in a controlled manner without having a negative impact on your external dependents.

18.1 Forward and Backward Compatibility

Compatibility works in two directions – forward and backward. When you update a component in such a way that other code that used the original version can continue to work with the updated version, you have made a *backward-compatible* change. When software that works with the updated version of your component also works with the original version, the changes are said to be *forward-compatible*.

Let's look at an example: a software house releases a library (v1.0) and an application, A1. Later, it re-releases the library (v2.0) and also releases a second application, A2, which uses it. If original copies of A1 continue to work as they did previously with v2.0 of the library, the changes made were **backward-compatible**.

If the v1.0 version of the library is then reinstalled for some reason, and A2 continues to work with the older library as it did previously with v2.0, the changes were also **forward-compatible**. This is illustrated in Figure 18.1.

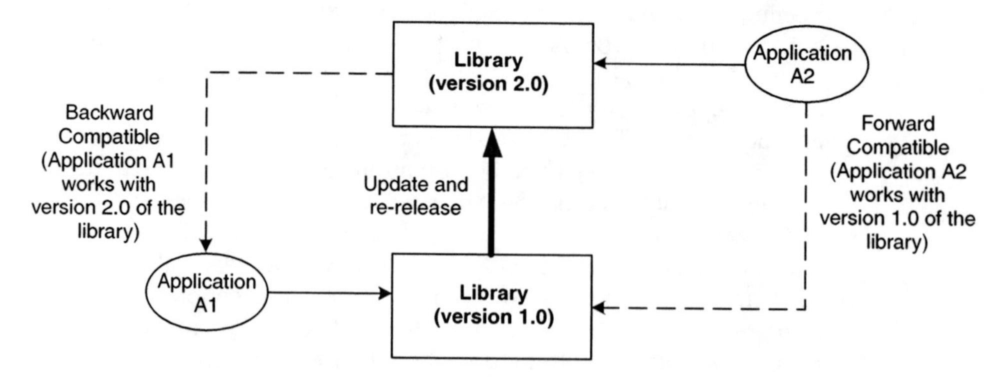

Figure 18.1 Forward and backward compatibility

Backward compatibility is usually the primary goal when making incremental releases of a component, with forward compatibility a desirable extra. Some changes cannot ever be truly forward-compatible, such as bug fixes, which by their nature do not work "correctly" in releases prior to the fix.

There are different levels of compatibility: class-level and library-level. Maintaining compatibility at a class level means ensuring that, among other things, methods continue to have the same semantics as were initially documented, no publicly accessible data is moved, and the size of an object of the class does not change (at least, not when a client is responsible for allocating memory for it). To maintain library-level compatibility, you should ensure that the API functions exported by a DLL are at the same ordinal and that the parameters and return values of each are still compatible.

There are also degrees to which compatibility can be maintained. This can vary from not requiring anything of dependent code (i.e. it can run as it did previously when a new version of the component it uses is installed) to requiring it to be recompiled and/or re-linked. However, if dependent source code needs to be changed in any way in order to work with a new release of your component, the changes were **incompatible**. Pure and simple.

Maintaining backward compatibility is the primary goal of an incremental component release.

18.2 Source Compatibility

If you make a change to your component which forces any dependent code to be changed in order to recompile, the change is **source-incompatible**.

```
// orchestra.h
// Version 1.0
...
class CInstrument : public CBase
    {
public:
    IMPORT_C static CInstrument* NewL();
    IMPORT_C virtual ~CInstrument();
public:
    IMPORT_C virtual void PlayL();
protected:
    CInstrument();
    void ConstructL();
private:
    // ...
    };
```

Thus, the following change is source-incompatible if dependent code includes `orchestra.h` and uses class `CInstrument`:

```
// orchestra.h
// Version 2.0, renames a class
...
class CStringInstrument : public CBase
    {
public:
    IMPORT_C static CStringInstrument* NewL();
    IMPORT_C virtual ~CStringInstrument();
public:
    IMPORT_C virtual void PlayL();
protected:
    CStringInstrument();
    void ConstructL();
private:
    // ...
    };
```

The only change between version 1.0 and version 2.0 is the change of class name from `CInstrument` to `CStringInstrument`. Any dependent binaries that built against version 1.0 and do not need to be recompiled will continue to work. It is only when the client code comes to be rebuilt against version 2.0 that it will be affected by source incompatibility (it must be modified to replace the `CInstrument` class with the new `CStringInstrument` class).

On Symbian OS, a typical source-incompatible change involves modifying the internals of a member function to give it the potential to leave when previously it could not do so. To adhere strictly to the naming convention, the name of the function must also be modified by the addition of a suffix `L`.

If you change your component and dependent components can recompile against the new version without the need to make any changes, it can be said to be **source-compatible**. An example of a source-compatible change is a bug fix to the internals of an exported function, which doesn't require a change to the function definition itself.

18.3 Binary Compatibility

Binary compatibility is achieved when one component, dependent on another, can continue to run without recompilation or re-linking after the component on which it depends is modified. The compatibility extends across compilation and link boundaries.

One example of a binary-compatible change is the addition to a class of a public, non-virtual function which is exported from the library with an ordinal[1] that comes after the previous set of exported functions. A

[1] As I described in Chapter 13, Symbian OS uses dynamic linking by ordinal, rather than function name matching, to reduce the size overhead of its export tables. However,

client component which was dependent on the original version of the library is not affected by the addition of a function to the end of the export list of the library. Thus the change can be said to be binary-compatible and backward-compatible (as well as source-compatible).

However, if the addition of a new function to the class causes the ordinals of the exported functions to be reordered, the change is not binary-compatible, although it continues to be source-compatible. Dependent code must be recompiled, because otherwise it would use the original, now invalid, ordinal numbers to identify the exports.

Of course, when we discuss the need for compatibility, this is with relation to the API which has been published to external components. You may make as many internally-incompatible changes to your code as you need to, since it is only your components that are affected. The restrictions I'll discuss in the rest of this chapter apply to changes which affect published, public APIs.

Let's move on to consider some of the practical aspects of code, to help you understand the kind of source code changes that can affect compatibility. I'll list some guidelines about which code changes do and do not affect the binary layout at a class or DLL level. These derive from an understanding of the way a compiler interprets the C++ standard. It is only by deriving a set of rules in this way that we can be confident as to which changes can be safely applied without breaking dependent code.

Many of the following guidelines are based on whether class methods or member data are "visible" to dependent code. This assumes that calling code gains visibility through legitimate means, that is, through exported header files, using standard object-oriented C++ to access and manipulate the methods and data encapsulated in C++ objects.

18.4 Preventing Compatibility Breaks

Do Not Change the Size of a Class Object

You should not change the size of a class object (e.g. by adding or removing data) unless you can guarantee that:

- your class is not externally derivable
- the only code that allocates an object resides within your component, or it has a non-public constructor that prevents it from being created on the stack

---

this has significant implications for binary compatibility, because the ordinals of exported functions must not be changed between one release of a DLL and another (otherwise code which originally used the old DLL will not be able to locate the functions it needs in the new version of the DLL).

- the class has a virtual destructor.

The reasons for this rule are fairly straightforward. First, the size of memory required for an object to be allocated on the stack is determined for each component at build time, and to change the size of an object would affect previously compiled client code, unless the client is guaranteed to instantiate the object only on the heap, say by using a `NewL()` factory function.

Additionally, access to data members within an object occurs through an offset from the `this` pointer. If the class size is changed, say by adding a data member, the offsets of the data members of derived classes are rendered invalid.

To ensure that an object of a class cannot be derived or instantiated except by members (or friends) of the class, it should have private, non-inline and non-exported constructors. It is not sufficient simply not to declare any constructors for the class, since the compiler will then generate an implicit, public default constructor. On Symbian OS, all C classes derive from `CBase`, which defines a protected default constructor, and prevents the compiler from generating an implicit version.[2] If your class needs a default constructor, you should simply define it as private and implement it in source, or at least, not inline where it is publicly accessible.

Here's a definition for a typical, non-derivable class:

```
Class CNonDerivable : public CBase
    {
public:
    // Factory - instantiates the object via ConstructL()
    IMPORT_C static CNonDerivable* NewL();
    // Virtual destructor, inherited from CBase
    IMPORT_C virtual ~CNonDerivable();
public:
    ... // Omitted for clarity - public exports & virtual functions
private:
    // constructor is private, not protected, to prevent subclassing
    CNonDerivable();
    void ConstructL();
private:
    ... // Omitted for clarity - private member data goes here
    };
```

So, to recap, unless you can guarantee that all objects have been constructed within your component on the heap, and are not derivable,

[2] Class `CBase` also declares, but does not define, a private copy constructor (and corresponding assignment operator) to prevent the compiler generating the implicit versions that perform potentially unsafe shallow copies of member data. Neither the copy constructor nor assignment operator are implemented, in order to prevent copy operations of `CBase` classes – although of course you can implement your own, public, versions for derived classes where this is valid.

you should not change the size of the member data of a class. If you follow these rules, you must also take into account that the object may be destroyed by an external component, which must deallocate all the memory it occupies. To guarantee that the object and the memory allocated for it are destroyed correctly, the class must have a virtual destructor. Again, you'll inherit a virtual destructor from `CBase` when implementing a C class.

You are unlikely to have a valid reason for preventing a stack-based T class from being instantiated on the stack, so you should never modify the size of an externally visible T class.

Do Not Remove Anything Accessible

If you remove something from an API which is used by an external component, that component's code will no longer compile against the API (a break in source compatibility) nor run against its implementation (a break in binary compatibility).

Thus, at an API level, you should not remove any externally-visible class, function, enumeration (or any value within an enumeration) or global data, such as string literals or constants.

At a class level, you should not remove any methods or class member data. Even private member data should not be removed, in general, because the size of the resulting object will change, which should be avoided for reasons described above.

Do Not Rearrange Accessible Class Member Data

If you don't remove class member data but merely rearrange it, you can still cause problems to client code that accesses that data directly because, as I described above, the offset of the member data from the object's `this` pointer will be changed.

You must not change the position of member data in a class if that data is:

- public (or protected, if the client can derive from the class)
- exposed through public or protected inline methods (which will have been compiled into client code).

This rule also means that you cannot change the order of the base classes from which a class multiply inherits without breaking compatibility, because this order affects the overall data layout of the derived object.

Do Not Rearrange the Ordinal of Exported Functions

In Chapter 13, I briefly mentioned the mechanism by which the API of a component is accessible to an external component, via a numerically ordered list of exported functions which are "frozen" into a module definition (`.def`) file. Each exported function is associated with an ordinal number, which is used by the linker to identify the function.

If you later re-order the `.def` file, say by adding a new export within the list of current exports, the ordinal number values will change and previously compiled code will be unable to locate the correct function. For example, say component A exports a method `DoSomethingL()` at ordinal 9, which is used by component B. Component A is later extended to export a new function, which is simply added to the start of the ordinal list, at ordinal 1. This changes the ordinal value of `DoSomethingL()` to 10. Unless component B is re-linked, it will still expect to be able to call `DoSomethingL()` at ordinal 9 but instead will call a completely different function. The change breaks binary compatibility. To avoid this, the new export should be added to the end of the `.def` file, which assigns it a new, previously unused, ordinal value.

The use of a `.def` file means that exported functions within a C++ class definition may be reordered safely, say to group them alphabetically or logically, as long as the underlying order of those exports in the `.def` file is not affected.

Do Not Add, Remove or Modify the Virtual Functions of Externally Derivable Classes

If a class is externally derivable (i.e. if it has an exported or inlined, public or protected constructor) you must not alter the nature of the virtual functions in any way, because this will break compatibility with calling code.

The reason why virtual functions should not be added to or removed from a derivable class is as follows: if the class is externally derivable and a derived class defines its own virtual functions, these will be placed in the virtual function table (`vtable`) directly after those defined by the base class, i.e. *your* class. If you add or remove a virtual function in the base class, you will change the `vtable` position of any virtual functions defined by a derived class. Any code that was compiled against the original version of the derived class will now be using an incorrect `vtable` layout.

This rule also applies to any base classes from which the class derives and the order from which they are derived if the class uses multiple inheritance. Changes to either will affect the layout of the `vtable`.

Likewise, you must not modify a virtual function if this means changing the parameters, the return type or the use of `const`. However, you can make changes to the internal operation of the function, for example a bug fix, as long as the documented behavior of the function is not changed.

Do Not Re-Order Virtual Functions

Although not stated in the C++ standard, the order in which virtual member functions are specified in the class definition can be assumed to be the only factor which affects the order in which they appear in the virtual function table. You should not change this order, since client code compiled against an earlier version of the virtual function table will call what has become a completely different virtual function.

Do Not Override a Virtual Function that was Previously Inherited

If you override a virtual function which was previously inherited, you are altering the virtual function table of the class. However, existing client code is compiled against the original `vtable` and will thus continue to access the inherited base-class function rather than the new, overridden, version. This leads to inconsistency between callers compiled against the original version of the library and those compiled against the new version. Although it does not strictly result in incompatibility, this is best avoided.

For example, a client of `CSiamese` version 1.0 calling `SleepL()` invokes `CCat::SleepL()`, while clients of version 2.0 invoke `CSiamese::SleepL()`:

```
class CCat : public CBase // Abstract base class
   {
public:
   IMPORT_C virtual ~CCat() =0;
public:
   IMPORT_C virtual void PlayL();  // Default implementation
   IMPORT_C virtual void SleepL(); // Default implementation
protected:
   CCat();
   };

class CSiamese : public CCat   // Version 1.0
   {
public:
   IMPORT_C virtual ~CSiamese();
public:
   // Overrides PlayL() but defaults to CCat::SleepL()
   IMPORT_C virtual void PlayL();
   // ...
   };

class CSiamese : public CCat   // Version 2.0
   {
public:
   IMPORT_C virtual ~CSiamese();
public:
   // Now overrides PlayL() and SleepL()
   IMPORT_C virtual void PlayL();
   IMPORT_C virtual void SleepL();
   // ...
   };
```

Do Not Modify the Documented Semantics of an API

If you change the documented behavior of a class or global function, or the meaning of a constant, you may break compatibility with previously published versions used by client code, regardless of whether source and binary compatibility are maintained. As a very simple example, you may supply a class which, when supplied with a data set, returns the average value of that data. If the first release of the `Average()` function returns the arithmetic mean value, the second release of `Average()` should not be changed to return the median value, or some other interpretation of an average. As always, of course, if all the callers of the function can accept the change, or if the effect is limited to the internals of your own components, then such a modification is acceptable.

Do Not Remove `const`

The semantics of "`const`" should not be removed, since this will be a source-incompatible change. This means that you should not remove the `const`-ness of a parameter, return type or class method that were originally specified as `const`.

Do Not Change from Pass by Value to Pass by Reference, or Vice versa

If parameters or return values were originally passed by value, you will break binary compatibility if you change them to reference values (or vice versa). When a parameter is passed by value to a function, the compiler generates a stack copy of the entire parameter and passes it to the function. However, if the function signature is changed to accept the parameter by reference, a word-sized reference to the original object is passed to the function instead. The stack frame for a pass-by-reference function call is thus significantly different from that for a pass-by-value function call. In addition, the semantics of passing by value and by reference are very different – as discussed in Chapter 20 – which inevitably causes binary incompatibility.

```
class TColor
    {
    ...
private:
    TInt iRed;
    TInt iGreen;
    TInt iBlue;
    };

// version 1.0
// Pass in TColor by value (12 bytes)
IMPORT_C void Fill(TColor aBackground);

// version 2.0 - binary compatibility is broken
```

```
// Pass in TColor by reference (4 bytes)
IMPORT_C void Fill(TColor& aBackground);
```

18.5 What Can I Change Without Breaking Binary Compatibility?

You Can Extend an API

You can add classes, constants, global data or functions without breaking compatibility.[3] Likewise, a class can be extended by the addition of static member functions or non-virtual member functions. Recall the definition of the direction of compatibility at the beginning of this chapter – any extension of an API is, by its very nature, not forward-compatible.

If you add functions that are exported for external code to use (as described in Chapter 20), you must add the new exports to the bottom of the module definition file (`.def`) used to determine the ordinal number by which the linker identifies the exported function. You must avoid reordering the list of exported functions. This will break binary compatibility, as described above.

As I discussed earlier, you should not add virtual member functions to classes which may have been subclassed (i.e., externally derivable classes), since this causes the `vtable` of the deriving classes to be re-ordered.

You Can Modify the Private Internals of a Class

Changes to private class methods that are not exported and are not virtual do not break client compatibility. Likewise for protected methods in a class which is not derivable. However, the functions must not be called by externally-accessible inline methods, since the call inside the inline method would be compiled into external calling code – and would be broken by an incompatible change to the internals of the class. As I discuss in Chapter 21, it is general good practice to restrict, or eliminate, the use of publicly-accessible inline functions, particularly where compatibility is an issue. This is also discussed further in Section 18.6.

Changes to private class member data are also permissible, unless it results in a change to the size of the class object or moves the position of public or protected data in the class object (exposed directly, through inheritance or through public inline "accessor" methods).

You Can Relax Access Specification

The C++ access specifier (`public`, `protected`, `private`) doesn't affect the layout of a class and can be relaxed without affecting the data order

[3] You should endeavor to avoid any name clashes within the component or others in the global namespace – otherwise the change will be source-incompatible.

of the object. The position of member data in a class object is determined solely by the position of the data as it is specified in the class definition.

It is not sensible to change the access specification to a more restricted form, e.g. from `public` to `private`, because, although it does not affect the code, it means that the member data becomes invisible to external clients when previously it was visible. A future change to the component might incorrectly assume that the data has always been private and modify its purpose, affecting external components dependent upon it. In addition, any existing client which accesses that data will no longer be able to do so – a source-incompatible change.

You Can Substitute Pointers for References and Vice Versa

Changing from a pointer to a reference parameter or return type (or vice versa) in a class method does not break binary compatibility. This is because references and pointers can be considered to be represented in the same way by the C++ compiler.

You Can Change the Names of Exported Non-Virtual Functions

Symbian OS is linked purely by ordinal and not by name and signature. This means that it is possible to make changes to the name of exported functions and retain binary, if not source, compatibility.

You Can Widen the Input

Input can be made more generic ("widened") as long as input that is currently valid retains the same interpretation. Thus, for example, a function can be modified to accept a less derived pointer[4] or extra values can be added to an enumeration, as long as it is extended rather than re-ordered, which would change the original values.

You Can Narrow the Output

Output can be made less generic ("narrowed") as long as any current output values are preserved. For example, the return pointer of a function can be made more derived as long as the new return type applies to the original return value.[5] For multiple inheritance, say, a pointer to a class is unchanged when it is converted to a pointer to the first base class in

---

[4] Say class `CSiamese` derives from class `CCat`. If the pointer passed to a function was originally of type `CSiamese`, it is acceptable to change the function signature to take a pointer to the less-derived `CCat` type.

[5] Using the same example of class `CSiamese` which derives from `CCat`, if the pointer returned from a function was originally of type `CCat`, it is acceptable to change the function signature to return a pointer to the more-derived `CSiamese` type.

the inheritance declaration order. That is, the layout of the object follows the inheritance order specified.

You Can Apply the `const` Specifier

It is acceptable to change non-`const` parameters, return types or the "`this`" pointer to be `const` in a non-virtual function, as long as the parameter is no more complicated than a reference or pointer (i.e., not a reference to a pointer or a pointer to a pointer). This is because you can pass non-`const` parameters to `const` functions or those that take `const` parameters. In effect, this is an extension of the "widen input" guideline described above.

18.6 Best Practice: Planning for Future Changes

Don't Inline Functions

As described earlier, and later in Chapter 21 which discusses good coding practice, an inline function is compiled into the client's code. This means that a client must recompile its code in order to pick up a change to an inline function.

If you want to use private inline methods within the class, you should ensure that they are not accessible externally, say by implementing them in a file that is accessible only to the code module in question.

Using an inline function increases the coupling between your component and its dependents, which should generally be avoided.

Don't Expose Any Public or Protected Member Data

The position of data is fixed for the lifetime of the object if it is externally accessible, either directly or through derivation. You should aim to encapsulate member data privately within a class and provide (non-inline) accessor functions where necessary. Chapter 20 discusses this in more detail.

Allow Object Initialization to Leave

The steps required to instantiate an object of your class may not currently have the potential to fail, but the class may be extended in future, perhaps to read data from a configuration file which may not succeed, say if the file is corrupt or missing. To allow for an extension of this type, object instantiation should be able to leave safely should this become necessary.

This is a straightforward guideline to adhere to, because the Symbian OS "two-phase construction" idiom, described in Chapter 4, allows for exactly this type of extensibility:

```
class CExtensibleClass : public CBase
    {
public:
    static CExtensibleClass* NewL();
    ... // Omitted
protected:
    CExtensibleClass();    // Constructor is externally inaccessible
    void ConstructL(){};   // For extensibility
private:
    ... // Omitted
    };

CExtensibleClass::CExtensibleClass()
    {// Implement non-leaving construction code in this function }

CExtensibleClass* CExtensibleClass::NewL()
    {// Can later be extended to call ConstructL() if initialization
     // needs to leave
    CExtensibleClass me = new (ELeave) CExtensibleClass();
    return (me);
    }
```

Override Virtual Functions that You Expect to Override Later

This will ensure that compatibility is not broken should you later need to override a virtual function which was originally inherited (as described earlier in Section 18.4). If you do not currently need to modify the functions beyond what the base class supplies, the overridden implementation should simply call the base-class function. This will allow the functions to be extended in future. Earlier, I gave the example of `CSiamese`, deriving from `CCat`, which inherited the default implementation of the `CCat::SleepL()` virtual method in version 1.0 but overrode it in version 2.0. The sample code below shows how to avoid this by overriding both virtual functions in version 1.0, although `CSiamese::SleepL()` simply calls through to `CCat::SleepL()`:

```
Class CCat : public CBase // Abstract base class
    {
public:
    IMPORT_C virtual ~CCat() =0;
public:
    IMPORT_C virtual void EatL();  // Default implementation
    IMPORT_C virtual void SleepL();// Default implementation
    // ...
    };

class CSiamese : public CCat    // Version 1.0
    {
```

```
    IMPORT_C virtual ~CSiamese();
public:
    IMPORT_C virtual void EatL(); // Overrides base class functions
    IMPORT_C virtual void SleepL();
    // ...
    };

// Function definitions not relevant to the discussion have been
// omitted
void CSiamese::EatL()
    {// Overrides base class implementation
    ... // Omitted for clarity
    }
void CSiamese::SleepL()
    {// Calls base class implementation
    CCat::SleepL();
    }
```

Provide "Spare" Member Data and Virtual Functions

As you'll have gathered from a number of the guidelines above, the use of virtual functions, although powerful in a C++ sense, can be quite limited in terms of extensibility when compatibility must be maintained. If it is possible that your class may need to be extended (and this is often hard to determine, so you may prefer to assume that it will), then you should add at least one reserve exported virtual function. This will give you the means by which to extend the class without disrupting the `vtable` layout of classes which derive from the class. In effect, this provides a means to get around the limits by which virtual functions may be extended, through the use of explicit interface design. You should, as always, implement any reserved functions, defaulting them to perform no action.

By the same token, you may wish to reserve at least four extra bytes of private member data in classes which may later need to be extended. This reserved data can be used as a pointer to extra data as it is required. Of course, if the class is internal to your component, or is unlikely to require later modification, you should not reserve extra memory in the class object, in order to continue to minimize the use of limited memory resources.

18.7 Compatibility and the Symbian OS Class Types

Chapter 1 describes the main class types, and their characteristics, on Symbian OS. The discussion in this chapter has mainly focused on compatibility issues which apply to C class objects and, to a lesser extent, R classes. T classes, by their very nature, tend to be stack-based and are often simple, taking a similar role to a C `struct`. T classes frequently do not have constructors and never have destructors. Once a T class is publicly defined, it is often difficult to make client-compatible changes to it.

18.8 Summary

A change is acceptable if every line of code affected by it can be altered, where necessary, and the code rebuilt against the changes. In effect, this often means that a change must be restricted to the internals of a component rather than its public API. For this reason, you should endeavor to keep private definitions and header files (and anything else which is likely to be subject to change) out of the public domain.

A change is also acceptable if it can be verified as compatible, according to the guidelines I've discussed here. In general, the key compatibility issue for shared library DLLs is backward binary compatibility, with source compatibility an additional aim.

These guidelines derive from those used within Symbian, in accordance with the C++ standard. I am very grateful to David Batchelor, John Forrest and Andrew Thoelke of Symbian who have previously written guides to compatibility best practice on Symbian OS, which I used as the basis of this chapter.

19

Thin Templates

Now, now, my good man, this is no time to be making enemies
Voltaire on his deathbed, in response to a priest who asked him to renounce Satan

C++ templates are useful for code that can be reused with different types, for example to implement container classes such as dynamic arrays. Templates allow the code to be generic, accepting any type, without forcing the programmer to overload a function.

```
template<class T>
class CDynamicArray : public CBase
    {
public:
    ... // Functions omitted for clarity
    void Add(const T& aEntry);
    T& operator[](TInt aIndex);
    };
```

Prior to the introduction of templates to the C++ standard, generic code tended to be written using `void*` arguments, to allow the caller to specify any pointer type as an argument. However, the major benefit of using C++ templates for genericity, instead of `void*`, is that templated code can be checked for type safety at compile time.

However, the problem with template code is that it can lead to a major increase in code size, because for each type used in the template, separate code is generated for each templated function. For example, if your code used the `CDynamicArray` class above for both an array of `HBufC*` and an array of `TUid` values, the object code generated when your code was compiled would contain two copies of the `Add()` function and two copies of `operator[]`, one for an array of `HBufC*` and one for an array of `TUid`. What's more, the template code is generated, at best, once for each DLL or EXE that uses it and, at worst, for every compilation unit. Compiling a templated class into a binary can thus have a significant impact on its size.

The advantage of automatically generated template code is thus a disadvantage when code size matters. Because Symbian OS code is deployed on mobile phones with limited ROM and RAM sizes, it is important to avoid code bloat. Using templated classes, such as `CDynamicArray` above, expands the code size too significantly. To avoid this, but still reap the benefits of C++ templates, Symbian OS makes use of the thin template pattern. This chapter describes the pattern in more detail, so you have a good idea of how the system code works. If you intend to use C++ templates[1] in your code on Symbian OS, you should endeavor to use this pattern to minimize code size. Let's consider the theory first, and then look at a couple of examples of the use of thin templates in Symbian OS code.

The thin template idiom works by implementing the necessary code logic in a generic base class that uses type-unsafe `TAny*` pointers rather than a templated type. This code is liable to misuse because of the lack of type-checking, so typically these methods will be made protected in the base class to prevent them being called naively. A templated class will then be defined which uses private inheritance[2] from this class to take advantage of the generic implementation. The derived class declares the interface to be used by its clients and implements it inline in terms of the base class. I'll show an example of this from Symbian OS code shortly.

Since the derived class uses templates, it can be used with any type required. It is type-safe because it is templated and thus picks up the use of an incorrect type at compile time. There are no additional runtime costs because the interfaces are defined inline, so it will be as if the caller had used the base class directly. And importantly, since the base class is not templated, only a single copy of the code is generated, regardless of the number of types that are used.

For illustration purposes, here is just a small part of the `RArrayBase` class and its subclass `RArray` (from `e32std.h` and `e32std.inl`). The type-unsafe base class (`RArrayBase`) implements the code logic for the array but cannot be used directly because all its methods are protected. You'll find a detailed discussion of the `RArray` class, and other Symbian OS container classes, in Chapter 7.

---

[1] You will typically want to use templates when writing a class that manipulates several different types using the same generic code. The code should be agnostic about the type passed into the template parameter, that is, the underlying logic is independent of type. Typical examples of (thin) template classes in Symbian OS are the array classes (I'll discuss `RArray` shortly), the singly- and doubly-linked list classes (based on `TSglQueBase` and `TDblQueBase`) and the circular buffers (based on `CCirBufBase`).

[2] Private inheritance means that the derived class is *implemented in terms of* the base class. Private inheritance is used when the deriving class uses some of the implemented methods of the base class, but has no direct conceptual relationship with the base class. Using private inheritance allows *implementation* to be inherited but all the methods of the base class become private members of the deriving class. In effect, the deriving class *does not inherit the interface of the base class.*

```
class RArrayBase
    {
protected:
    IMPORT_C RArrayBase(TInt anEntrySize);
    IMPORT_C RArrayBase(TInt aEntrySize,TAny* aEntries, TInt aCount);
    IMPORT_C TAny* At(TInt anIndex) const;
    IMPORT_C TInt Append(const TAny* anEntry);
    IMPORT_C TInt Insert(const TAny* anEntry, TInt aPos);
    ...
    };
```

The templated `RArray` class privately inherits the implementation and defines a clear, usable API for clients. The API is defined inline and uses the base class implementation. Elements of the array are instances of the template class.

```
template <class T>
    class RArray : private RArrayBase
    {
public:
    ...
    inline RArray();
    inline const T& operator[](TInt anIndex) const;
    inline T& operator[](TInt anIndex);
    inline TInt Append(const T& anEntry);
    inline TInt Insert(const T& anEntry, TInt aPos);
    ...
    };

template <class T>
inline RArray<T>::RArray()
    : RArrayBase(sizeof(T))
    {}
template <class T>
inline const T& RArray<T>::operator[](TInt anIndex) const
    {return *(const T*)At(anIndex); }
template <class T>
inline T& RArray<T>::operator[](TInt anIndex)
    {return *(T*)At(anIndex); }
template <class T>
inline TInt RArray<T>::Append(const T& anEntry)
    {return RArrayBase::Append(&anEntry);}
template <class T>
inline TInt RArray<T>::Insert(const T& anEntry, TInt aPos)
    {return RArrayBase::Insert(&anEntry,aPos);}
```

Use of the class is then straightforward:

```
void TestRArray()
    {
    const TInt arraySize = 3;
    RArray<TInt> myArray(arraySize);
    for (TInt index = 0; index<arraySize; index++)
        {
```

```
        myArray.Append(index);
        }

    TInt count = myArray.Count();
    ASSERT(arraySize==count);
    for (index = 0; index<arraySize; index++)
        {
        ASSERT(myArray[index]==index);
        }
    }
```

For another example of the thin template pattern in Symbian OS, consider the `TBufC` and `TBuf` descriptor classes discussed in Chapter 5. These classes are templated on an integer, the value of which is used as the variable which determines the maximum statically-allocated length of the buffer. In this case, the inheritance model is public and the derived class publicly inherits the base class implementation (whereas the previous example used private inheritance to gain access to the implementation without inheriting the behavior).

The constructors of the derived `TBufC16` class, and other functions that use the template parameter, are declared inline. I've shown the constructors for the base and derived classes below (from `e32des16.h` and `e32std.inl`):

```
class TBufCBase16 : public TDesC16
    {
protected:
    IMPORT_C TBufCBase16();
    inline TBufCBase16(TInt aLength);
    IMPORT_C TBufCBase16(const TUint16 *aString,TInt aMaxLength);
    IMPORT_C TBufCBase16(const TDesC16 &aDes,TInt aMaxLength);
    ...
    };

template <TInt S>
class TBufC16 : public TBufCBase16
    {
public:
    inline TBufC16();
    inline TBufC16(const TUint16 *aString);
    inline TBufC16(const TDesC16 &aDes);
    ...
    };

template <TInt S>
inline TBufC16<S>::TBufC16()
    : TBufCBase16()
    {}
template <TInt S>
inline TBufC16<S>::TBufC16(const TUint16 *aString)
    : TBufCBase16(aString,S)
    {}
template <TInt S>
inline TBufC16<S>::TBufC16(const TDesC16 &aDes)
```

```
: TBufCBase16(aDes,S)
{}
```

To get the benefits of C++ templates without the disadvantages of code bloat, you should prefer the thin-template idiom which is used throughout Symbian OS code.

19.1 Summary

This chapter explained why C++ templates are ideal for reusable type-safe but type-agnostic code, but have the disadvantage that they can increase their clients' code size quite considerably. For each templated class, every time a different type is used, separate code is generated for every templated function. Furthermore, this code duplication occurs in each client DLL or compilation unit using the templated class. This can cause significant code bloat unless the number of different types that are used with the templated class is limited, the code generated is small, and it is guaranteed that only a few clients will use the templated class.

Many Symbian OS container classes use the thin template pattern to gain the advantages of C++ templates without the disadvantages of increased code size (which is unacceptable on the "small footprint" devices on which Symbian OS is deployed).

This chapter described the characteristics of the thin template pattern, which typically defines a base class (containing a generic implementation, usually specified as `protected` to prevent it being called directly) and a derived class (which uses private inheritance to inherit the implementation of the base class). The derived class exposes a templated interface, implemented inline in terms of the base class and thus benefits from compile-time type-checking by using C++ templates. The derived class does not have the associated size overhead because, regardless of the number of types used, the code is inline and uses a single copy of the generic base class implementation.

In this chapter the `RArray`, `TBuf` and `TBufC` classes illustrated the thin template idiom as it is employed on Symbian OS, and how these classes should be used. The `RArray` class is discussed in detail in Chapter 7, while `TBuf` and `TBufC` are examined in Chapter 5.

20

Expose a Comprehensive and Comprehensible API

Have no fear of perfection – you'll never reach it
Salvador Dali

So, you're ready to design a class. You've read Chapter 1, and have thought about the characteristics of your class, maybe by asking the following questions: "How will objects of my class be instantiated?"; "Will they be stack- or heap-based?"; "Will the class have any data or simply provide an interface?"; "Will it need a destructor?" The answers to these questions will help you decide what kind of Symbian OS class it is and give you the first letter of its name, the rest of which is down to you. When choosing a class name it is frequently difficult to be descriptive yet brief. Unless your class consists solely of static functions, it will be instantiated and used as an object. The class name should be a noun to reflect that. I'll discuss the best strategy for naming classes and their member data, methods and parameters, later in this chapter.

The type of class you choose affects, to some extent, the definition of the class. For example, if you decide to implement it as a T class, it won't have a destructor nor own any data which needs one. On the other hand, if you're writing a C class, you'll most likely need to use two-phase construction. To do this, you will make constructors protected or private and provide a public static function, usually called `NewL()`, which calls a non-public second-phase construction method to perform any initialization that may leave. You'll find more information about two-phase construction in Chapter 4.

This chapter aims to highlight the factors you need to consider when designing a class, in particular the application programming interface (API) it will expose to its clients. The chapter isn't a comprehensive treatise in class design – that would need a whole book – but it does point out some of the more important points for good C++ class design on Symbian OS.

The quality of your API is important if your clients are to find it easy to understand and use. It should provide all the methods they need to perform the tasks for which the class is designed, but no more. You should design your interface to be as simple as possible, each member function having a distinct purpose and no two functions overlapping in what they perform. Try and limit the number of methods, too – if your class provides too many it can be confusing and difficult to work with. A powerful class definition has a set of functions where each has a clear purpose, making it straightforward, intuitive and attractive to reuse. The alternative might make a client think twice about using the class – if the functions are poorly declared, why trust the implementation? Besides the benefit of avoiding duplicating the implementation effort, with limited memory space on a typical Symbian OS phone, it's in everyone's interests to reuse code where possible.

Even if your own code is the intended client, no matter! Should you decide later to reuse the class, the time you spend making it convenient to use will pay off. By making the class simple to understand and use, and cutting out duplicated functionality, you'll cut down your test and maintenance time too. Furthermore, a well-defined class makes documentation easier and that's got to be a good thing.

20.1 Class Layout

Before looking at details of good API design, let's consider the aesthetics of your class definition. When defining a class, make it easy for your clients to find the information they need. The convention is to lay out Symbian OS classes with public methods at the top, then protected and private methods, following this with public data if there is any (and later in this chapter I'll describe why there usually shouldn't be), protected and private data. I tend to use the `public`, `protected` and `private` access specifiers more than once to split the class into logically related methods or data. It doesn't have any effect on the compiled C++, and it makes the class definition simpler to navigate. For example:

```
class CMyExample : public CSomeBase
    {
public: //  Object instantiation and destruction
    static CMyExample* NewL();
    static CMyExample* NewLC();
    ~CMyExample();
public:
    void PublicFunc1();          // Public functions, non virtual
    ...
public:
    inline TBool Inline1();      // Inline (defined elsewhere)
    ...
public:
```

```
    virtual void VirtualFunc1(); // Public virtual
    ...
protected:
    // Implementation of pure virtual base class functions
    void DoProcessL();
    ...
    // etc for other protected and private functions
private: // Construction methods, private for 2-phase construct
    void ConstructL();
    CMyExample();
    ...
private:               // Data can also be grouped, e.g. into that owned
    CPointer* iData1;  // by the class (which must be freed in the
    ...                // destructor) & that which does not need cleanup
    };
```

20.2 IMPORT_C and EXPORT_C

Having declared your class, you should consider how your clients get access to it. If it will be used by code running in a separate executable (that is, DLL or EXE code compiled into a separate binary component) from the one in which you will deliver the class, you need to export functions from it. This makes the API "public" to other modules by creating a `.lib` file, which contains the export table to be linked against by client code. There's more about statically linked DLLs in Chapter 13.

Every function you wish to export should be marked in the class definition in the header file with `IMPORT_C`. Admittedly, this is slightly confusing, until you consider that the client will be including the header file, so they are effectively "importing" your function definition into their code module. You should mark the corresponding function definition in the `.cpp` file with `EXPORT_C`. The number of functions marked `IMPORT_C` in the header files must match the number marked `EXPORT_C` in the source.

Here's an example of the use of the macros, which will doubtless look familiar from class definitions in the header files supplied by whichever Symbian OS SDK you use.

```
class CMyExample : public CSomeBase
  {
public:
  IMPORT_C static CMyExample* NewL();
public:
  IMPORT_C void Foo();
  ...
  };

EXPORT_C CMyExample* CMyExample::NewL()
{...}
```

```
EXPORT_C void CMyExample::Foo()
{...}
```

The rules as to which functions you need to export are quite simple. Firstly, you must never export an inline function. There's simply no need to do so! As I described, `IMPORT_C` and `EXPORT_C` add functions to the export table to make them accessible to components linking against the library. However, the code of an inline function is, by definition, already accessible to the client, since it is declared within the header file. The compiler interprets the inline directive by adding the code *directly* into the client code wherever it calls it. In fact, if you export an inline function, you force all DLLs which include the header file to export it too, because the compiler will generate an out-of-line copy of the function in every object file which uses it. You'll find a warning about the use of inline directives except for the most trivial of functions in Chapters 18 and 21.

Only functions which need to be used outside a DLL should be exported. When you use `IMPORT_C` and `EXPORT_C` on a function, it adds an entry to the export table. If the function is private to the class and can never be accessed by client code, exporting it merely adds it to the export table unnecessarily.

Private functions should only be exported if:

- they are virtual (I'll discuss when to export virtual functions shortly)
- they are called by a public inline function (this is clear when you consider that the body of the inline function will be added to client code, which means that, in effect, it makes a direct call to the private function).

Similarly, a protected function should only be exported if:

- it is called by an inline function, as described above
- it is designed to be called by a derived class, which may be implemented in another DLL
- it is virtual.

All virtual functions, public, protected or private, should be exported, since they may be re-implemented by a derived class in another code module. Any class which has virtual functions must also export a constructor, even if it is empty.

The one case where you should not export a virtual function is if it is pure virtual. This is obvious when you consider that there is generally no implementation code for a pure virtual function, so there is no code to export. As long as a deriving class has access to the virtual function table for your base class, it is aware of the pure virtual function and is forced to implement it. In the rare cases where a pure virtual function has a function body, it must be exported.

The virtual function table for a class is created and updated by the constructor of the base class and any intermediate classes. If you don't export a constructor, when a separate code module comes to inherit from your class, the derived constructor will be unable to access the default constructor of your class in order to generate the virtual function table. This is why any class which has virtual functions must export a constructor. Remember the rule that you must not export inline functions? Since you're exporting the constructor, you must implement it in your source module, even if it's empty, rather than inline it. A good example of this is class `CBase` (defined in `e32base.h`), which defines and exports a protected default constructor.

The rules of exporting functions are as follows:

- **never export an inline function**
- **only those non-virtual functions which need to be used outside a DLL should be exported**
- **private functions should only be exported if they are called by a public inline function**
- **protected functions should be exported if they are called by a public inline function or if they are likely to be called by a derived class which may be implemented in another code module**
- **all virtual functions, public, protected or private, should be exported, since they may be re-implemented in a separate module**
- **pure virtual functions should not be exported unless they contain code**
- **any class which has virtual functions must also export a (non-inlined) constructor, even if it is empty.**

20.3 Parameters and Return Values

Let's move on to think about the definitions of class methods in terms of parameters and return values. I'll compare passing and returning values by value, reference and pointer, and state a few general guidelines for good C++ practice on Symbian OS.

Pass and Return by Value

```
void PassByValue(TExample aParameter);
TExample ReturnAValue();
```

Passing a parameter by value, or returning a value, takes a copy of the argument or final return value. This can potentially be expensive, since it invokes the copy constructor for the object (and any objects it encapsulates) as well as using additional stack space for the copy. When the object passed in or returned goes out of scope, its destructor is called. This is certainly less efficient than passing a reference or pointer to the object as it currently exists, and you should avoid it for large or complex objects such as descriptors.[1] However, it is insignificant for small objects (say less than eight bytes) and the built-in types (`TBool`, `TText`, `TInt`, `TUint` and `TReal`). Of course, by taking a copy of the parameter, the original is left unchanged, so a parameter passed by value is most definitely a constant input-only parameter.

You may even choose to return an object by `const` value in some cases. This prevents cases of assignment to the return value – which is illegal for methods that return the built-in types. You should strive to make your own types behave like built-in types and may choose to use a `const` return value to enforce this behavior when you are returning an instance of your class by value.

Pass and Return by `const` Reference

```
void PassByConstRef(const TExample& aParameter);
const TExample& ReturnAConstRef();
```

Passing an object by `const` reference prevents the parameter from being modified and should be used for constant input parameters larger than the built-in types. Equally, for efficiency reasons, returning a value as a constant reference should be used in preference to taking a copy of a larger object to return it by value. You must be careful when returning an object by reference, whether it is modifiable or constant, because the caller of the function and the function itself must agree on the

[1] When passing objects by reference or pointer, a 32-bit pointer value is transferred. This minimal memory requirement is fixed, regardless of the type to which it points. Internally, a reference is implemented as a pointer, with additional syntax to remove the inconvenience of indirection.

lifetime of the object which is referenced. I'll discuss this further in the next section.

Pass and Return by Reference

```
void PassByRef(TExample& aParameter);
TExample& ReturnARef();
```

Passing an object by reference allows it to be modified; it is, in effect, an output or input/output parameter. This is useful for both larger objects and the built-in types, and is a common alternative to returning an object by value.

Returning a reference is commonly seen to allow read/write access, for example to a heap-based object for which ownership is not transferred. Returning a reference passes access to an existing object to the caller; of course, the lifetime of this object must extend beyond the scope of the function.

Consider the following example:

```
class TColor
    {
public:
    TColor(TInt aRed, TInt aGreen, TInt aBlue);
    TColor& AddColor(TInt aRed, TInt aGreen, TInt aBlue);
    //... functions omitted for clarity
private:
    TInt iRed;
    TInt iGreen;
    TInt iBlue;
    };

TColor::TColor(TInt aRed, TInt aGreen, TInt aBlue)
: iRed(aRed), iGreen(aGreen), iBlue(aBlue)
    {}

TColor& TColor::AddColor(TInt aRed, TInt aGreen, TInt aBlue)
    {// Incorrect implementation leads to undefined behavior
    TColor newColor(aRed+iRed, iGreen+aGreen, iBlue+aBlue);
    return (newColor); // Whoops, new color goes out of scope
    }

TColor prettyPink(250, 180, 250);
TColor differentColor = prettyPink.AddColor(5, 10, 5);
```

The `AddColor()` method returns a reference to `newColor`, a local stack-based object which ceases to exist outside the scope of `AddColor()`. The result is undefined, and some compilers flag this as an error. If your compiler does let you build and run the code, it may even succeed, since different compilers allow temporary variables different life spans. In circumstances like these, that's what "undefined" means. What works with one compiler will doubtless fail with another and you should avoid any reliance upon compiler-specific behavior.

The previous code is a classic example of knowing that it's generally preferable to return objects larger than the built-in types by reference rather than by value, but applying the rule where it is not actually valid to do so. An equally invalid solution would be to create the return value on the heap instead of the stack, to avoid it being de-scoped, and then return a reference to the heap object. For example, here's another incorrect implementation of adding two colors, this time in a leaving function.

```
TColor& TColor::AddColorL(TInt aRed, TInt aGreen, TInt aBlue)
    {// Incorrect implementation leads to a memory leak
    TColor* newColor = new (ELeave) TColor(aRed+iRed, iGreen+aGreen,
            iBlue+aBlue);
    return (*newColor);
    }
```

The `newColor` object does at least exist, but who is going to delete it? By returning a reference, the method gives no indication that the ownership of `newColor` is being transferred to the method's caller. Without clear documentation, the caller wouldn't know that they're supposed to delete the return value of `AddColorL()` when they've finished with it.

```
TColor prettyPink(250, 180, 250);
TColor differentColor = prettyPink.AddColor(5, 10, 5);
...
delete &differentColor; // Nasty. It's unusual to delete a reference!
```

This just isn't the done thing with functions which return objects by reference! Even if most of your callers read the documentation and don't object heartily to this kind of behavior, it only takes one less observant caller to leak memory.

Of course, there is an expense incurred in return by value, namely the copy constructor of the object returned and, later, its destructor. But in some cases, such as that described above, you have to accept that (and in some cases the copy constructor and destructor are trivial so it's not an issue). In addition, the compiler may be able to optimize the code (say, using the named return value optimization) so the price you pay for intuitive, memory-safe and well-behaved code is often a small one. Here's the final, correct implementation of `AddColor()`, which returns `TColor` by value:

```
TColor TColor::AddColor(TInt aRed, TInt aGreen, TInt aBlue)
    {
    TColor newColor(aRed+iRed, iGreen+aGreen, iBlue+aBlue);
    return (newColor); // Correct, return by value
    }
```

If you return a reference to an object which is local to the scope of your function, the object will be released back to the stack when your function returns, and the return value will not reference a valid object. In general, if an object doesn't exist when the function that returns it is called, you can only return it by reference if some other object, whose life-time extends beyond the end of that function, has ownership of it.

Pass and Return by `const` Pointer

```
void PassByConstPtr(const CExample* aParameter);
const CExample* ReturnAConstPtr();
```

Passing or returning a constant pointer is similar to using a constant reference in that no copy is taken and the recipient cannot change the object.[2] It's useful to return a pointer when returning access to a C++ array, or where an uninitialized parameter or return value is meaningful. There's no such thing as an uninitialized reference, so returning a NULL pointer is the only valid way of indicating "no object" in a return value. However, you might consider overloading a method to take an additional constant reference input parameter. This means that you can use a `const` reference for cases where the input parameter can be supplied, and call the overloaded method which doesn't take any input parameter when no input is supplied. For example:

```
// Take a const pointer (aInput may be NULL)
void StartL(const CClanger* aInput, const TDesC8& aSettings);

// Or alternatively overload to take a const reference when it is valid
// Use this overload in cases where aInput is NULL
void StartL(const TDesC8& aSettings);
// Use this overload in cases where aInput!=NULL
void StartL(const CClanger& aInput, const TDesC8& aSettings);
```

The benefit of using a reference over a pointer is that it is always guaranteed to refer to an object. In the first overload, `StartL()` would have to implement a check to see whether `aInput` points to an object before dereferencing it, in case it is NULL. This is unnecessary in the third overload, where `aInput` is guaranteed to refer to an object.

[2] The `const`ness of the pointer can, of course, be cast away using `const_cast<CExample*>(aParameter)`, but you wouldn't do that without good reason, would you?

Pass and Return by Pointer

```
void PassByPtr(CExample* aParameter);
CExample* ReturnAPtr();
```

Passing or returning a pointer allows the contents to be modified by the recipient. As with const pointers, you should pass or return a pointer if it can be NULL, if it points into a C++ array or if you are transferring ownership of an object to a caller. On Symbian OS, passing or returning a pointer often indicates a transfer of ownership and it's preferable not to return a pointer if you are not transferring ownership (since you must then document clearly to callers that they should not delete it when they've finished with it). In fact, you should prefer to pass by reference or return a reference rather than use pointers except for the reasons described.

All else being equal, a reference can be more efficient than a pointer when the compiler must maintain a NULL pointer through a conversion. Consider the following class and pseudo-code:

```
class CSoupDragon : public CBase, public MDragon
    {...};

CSoupDragon* soupDragon;
MDragon* dragon;
...
dragon = soupDragon; // soupDragon may be NULL
```

For the conversion between CSoupDragon and MDragon, the compiler must add sizeof(CBase) to the soupDragon pointer in all cases, except where soupDragon is NULL, whereupon it must continue to be NULL rather than point incorrectly at the address which is the equivalent of sizeof(CBase). Thus the compiler must effect the following:

```
dragon = (MDragon*)(soupDragon ? (TUint8*)soupDragon+sizeof(CBase) :
NULL);
```

For a conversion which involves references rather than pointers, the test is unnecessary, since a reference must always refer to an object.

For any of your code that receives a pointer return value, or if you implement methods that take pointer parameters, you should consider the implications of receiving a NULL pointer. If an uninitialized value is incorrect, i.e. a programming error, you should use an assertion statement to verify that it is valid. The use of assertions for "defensive" programming is discussed further in Chapter 16.

20.4 Member Data and Functional Abstraction

In this discussion on the API of your class, I've discussed the definition of the member functions of your class, but not really touched on the member data. There's a very simple rule, which is that you should not make class data public, and there's a good reason for this – encapsulation.

The benefit of keeping your member data private to the class is that you can control access to it. First, you can decide whether to expose the data at all; if you choose to do so, you can provide member functions in your class to expose the data. If you follow the guidelines above, you can control precisely the type of access allowed. If you return `const` references, `const` pointers or a value, the caller has read-only access, while returning a reference or a pointer allows the caller to modify the data, as illustrated by the following:

```
const TExample& ReadOnlyReference();
const TExample* ReadOnlyPointer();
TExample ReadOnlyValue();
TExample& ReadWriteReference();
TExample* ReadWritePointer();
```

An additional benefit of keeping class member data private is a degree of functional abstraction. By providing methods to `Set()` and `Get()`, the variable is not exposed directly. Should you later decide to change the implementation of the class, you can do so without requiring your clients to update their code.

For example, consider this rather contrived class which stores a password and compares it with a password typed in later, providing a re-usable class for applications to protect their user files. In version 1.0, the unencrypted password is, rather naively, stored within the object. The code is released and everyone is happy for a while, including a few hackers. A code review before version 2.0 highlights the problem. If the class has been declared as follows, any attempt to add more security to the class forces clients of the class to change their code. Incidentally, I discuss how to maintain compatibility for client code in Chapter 18.

```
// Version 1.0
class CPassword : public CBase
    {
public:
    ... // Other functions omitted for clarity
public:
    // Bad! There are no access functions - the member data is public
    // The caller can set and get the password directly
    HBufC8* iPassword;
    };
```

The same problem arises in a second definition (version 1.1) below, but for different reasons. This class is a step in the right direction, because the password data is at least private. To set and get the password, the caller must call a method explicitly rather than modify the contents of the object directly.

```
// Version 1.1
class CPassword : public CBase
    {
public:
    ...
    IMPORT_C void SetPasswordL(const TDesC8& aPassword);
    inline HBufC8* Password(); // Better! But not great
private:
    HBufC8* iPassword; // Better. Password is private
    };

inline HBufC8* CPassword::Password()
    {
    return (iPassword);
    }

// SetPasswordL() is a leaving method because allocation
// of iPassword may leave if there is insufficient memory
EXPORT_C void CPassword::SetPasswordL(const TDesC8& aPassword)
    {
    HBufC8* newPassword = aPassword.AllocL();
    delete iPassword;
    iPassword = newPassword;
    }
```

To access the data, the caller must now call `Password()`, which provides read/write access to `iPassword`. Sure, it returns a constant heap descriptor, but from Chapters 5 and 6, you know how to call `Des()` on that pointer to get a modifiable pointer which can be used to update the contents of the buffer. This is far from ideal; the password should only be modifiable through the `SetPasswordL()` method, otherwise it's confusing for the client to know which to use. To get read-only access, the method should either return a `const HBufC8*` or, preferably, return a constant reference to the more generic descriptor type, `TDesC8`, as follows:

```
// Version 1.2
class CPassword : public CBase
    {
public:
    ...
    // Better! Now a const method and return value
    inline const TDesC8& Password() const;
private:
```

```
    HBufC8* iPassword;
    };

inline const TDes8C& CPassword::Password() const
    {
    return (*iPassword);
    }
```

In fact, for access to the password, this class definition is not much of an improvement on the original one, because the method exposes the class implementation directly. It's questionable whether there's much benefit over the original version, since it doesn't do anything additional to version 1.0, other than requiring a function call to retrieve the password. In fact, by implementing the accessor in an inline method, the additional overhead of a function call is removed at compile time. This is not necessarily a good thing in this case because its implementation is compiled into the caller's code. It isn't possible to update `Password()` without forcing clients of the class to recompile to receive the benefit of the new implementation, as I discussed in Chapter 18.

Here's a better definition of the class (version 1.3). Notice that the data member is private; there is a single method to set the password and, rather than returning the password, a method is provided to compare an input password with the stored value, returning a `TBool` result to indicate a match. This method is not inlined:

```
// Version 1.3
class CPassword : public CBase
    {
public:
    IMPORT_C TBool ComparePassword(const TDesC8& aInput) const;
    // Implemented as shown previously
    IMPORT_C void SetPasswordL(const TDesC8& aPassword);
private:
    HBufC8* iPassword;
    };

EXPORT_C TBool CPassword::ComparePassword(const TDesC8& aInput) const
    {
    return (0==iPassword->Compare(aInput));
    }
```

A more secure storage mechanism is used in version 2.0, which doesn't require the definition of class `CPassword` to change, but updates the internals of `SetPasswordL()` and `ComparePassword()` to store the password as a hash value, rather than "in the clear". By abstracting the functionality into the class, when version 2.0 is released existing clients of the library are unaffected by the upgrade.

```
// Version 2.0
EXPORT_C void CPassword::SetPasswordL(const TDesC8& aPassword)
    {
    TBuf8<KMaxHashSize> hashed;
    // Fill hashed with a hash of aPassword using an appropriate method
    // ...
    HBufC8* newPassword = hashed.AllocL();
    delete iPassword;
    iPassword = newPassword;
    }

EXPORT_C TBool CPassword::ComparePassword(const TDesC8& aInput) const
    {
    TBuf8<KMaxHashSize> hashed;
    // Fill hashed with a hash of aInput using an appropriate method
    // ...
    return (0==iPassword->Compare(hashed));
    }
```

Of course, there are still problems with this class, not least the fact that there is no error checking to see whether `iPassword` has actually been set by a call to `SetPasswordL()`. A better implementation would perhaps remove `SetPasswordL()` and require the password to be passed to a `NewL()` or `NewLC()` two-phase construction method, as described in Chapter 4. This would also limit when, in the object's lifetime, the password could change, for better security. I'll leave that implementation as an exercise for the reader.

Where possible, use `const` on parameters, return values, "query" methods, and other methods that do not modify the internals of a class. It indicates invariance to other programmers and ensures that the compiler enforces it.

20.5 Choosing Class, Method and Parameter Names

The title of this chapter includes the word "comprehensible", and this is an important issue when it comes to naming your class and the methods of its API. In all cases, you should strive for clear and distinct names, without making them so long as to be burdensome or, worse, abbreviating each to an acronym of your own choosing. Likewise, the names of the parameters in each method should be clear and descriptive. If you can, make it possible for your clients to write comprehensible code too. For example, in methods that take a number of initialization values, it's a good idea to use custom enumerations with clear names. Not only does this make it easy for your client to work out which settings to choose, but it's easier when looking back at the code to understand what it means. For example, take the following function declaration in class `CGpSurgery`:

```
class CGpSurgery : public CBase
    {
public:
    void MakeHospitalAppointment(TBool aExistingPatient,
            TBool aUrgency, TBool aTestData);
    ... // Other functions omitted for clarity
    };
```

Without looking at the declaration of `MakeHospitalAppointment()`, it's not immediately clear how to call the method, for example to make an urgent hospital appointment for a new patient who already has some test data available. Unless you name your parameters carefully, your client may well have to consult your documentation too, to find out the appropriate boolean value for each variable.

```
MakeHospitalAppointment(EFalse, ETrue, ETrue);
```

A far clearer approach is to use enumeration values whose name clearly indicates their purpose. However, if you define them in global scope, you or your clients may well find that a name clash arises with code in other header files, particularly if you choose short, simple names such as `TSize` or `TColor`. Even if your code doesn't get a clash, this (ab)use of the global scope means that code which includes your header may get a conflict later down the line. It's preferable to define them inside the class to which they apply. The caller then uses the class scope to identify them. This is a useful approach for class-specific enumerations, typedefs and constants. If you prefer to keep them out of your class scope, you could alternatively use a C++ namespace to prevent spilling your definitions into the global scope.

```
class CGpSurgery : public CBase
    {
public:
    enum TPatientStatus { EExistingPatient, ENewPatient };
    enum TAppointmentUrgency { EUrgent, ERoutine };
    enum TTestData { ETestResultsPending, ETestResultsAvailable };
public:
    void MakeHospitalAppointment(TPatientStatus aExistingRecords,
            TAppointmentUrgency aUrgency, TTestData aTestData);
    };
```

Looking at the function call, it is now significantly clearer what each parameter refers to; in effect, the code has documented itself.

```
MakeHospitalAppointment(CGpSurgery::ENewPatient, CGpSurgery::EUrgent,
        CGpSurgery::ETestResultsAvailable);
```

The use of enumerations rather than boolean values also provides extensibility in the future; in the case of the `CGpSurgery` class, additional levels of appointment urgency, patient status or test data availability can be introduced without the need to change the signature of `MakeHospitalAppointment()`. This avoids breaking compatibility – which is the subject of Chapter 18.

20.6 Compiler-Generated Functions

I've discussed some of the things to consider when defining your class, but before concluding this chapter, it's worth briefly describing the functions that the compiler generates for you if you don't add them yourself. If you have not declared a copy constructor, assignment operator or destructor, the compiler generates them implicitly for a class in case they need to be invoked. If there are no constructors declared, it declares a default constructor too.

The implicitly-generated functions are public; the constructor and destructor are simply placeholders for the compiler to add the code required to create and destroy an object of the class (for example, to set up or destroy the virtual function table). The destructor does not perform any cleanup code. The compiler-generated copy constructor and assignment operator perform a copy or assignment on each member of your class, invoking the copy constructor or assignment operator if there is one defined, or applying the rule for the members of the encapsulated object if not. Built-in types, pointers and references are copied or assigned using a shallow bitwise copy. This is problematic for pointers to objects on the heap, since a bitwise copy is rarely desirable, opening up opportunities for dangling pointers; these can result in a memory leak or multiple deletion through the same pointer, which raises a panic (`USER 44`).

This is particularly true for `CBase`-derived classes which should be constructed through the `NewL()` and `NewLC()` functions, which are guaranteed to initialize the object correctly using two-phase construction (as described in Chapter 4). They should not be copy constructed or assigned to, because this bypasses the two-phase construction and makes shallow copies of any pointers to dynamic memory.

To prevent accidental copies, `CBase` declares a private copy constructor and assignment operator (as you'll see from the class definition in `e32base.h`). Declaring, but not defining, a private copy constructor and assignment operator prevents calling code from performing invalid copy operations using compiler-generated code. If your C class does need a copy constructor, you must explicitly declare and define one publicly – or provide a `CloneL()` or `CopyL()` method, which allows you to make leaving calls, which are not possible in a copy constructor.

In fact, you'll probably not find yourself implementing a copy constructor or assignment operator on Symbian OS very often. As I've explained, C class objects are not suitable for copy or assignment, while T class objects are ideal candidates for allowing the compiler to generate its own implicit versions. An R class object is unlikely to be copied and is safe for bitwise copy anyway. You would usually not expect to define an explicit copy constructor or assignment operator for an R class unless a shallow copy of the resource handle causes problems. For example, while the first object to call `Close()` releases the underlying resource and zeroes the handle value (making it invalid), a second object may still have a non-zero handle value. If the second object attempts to use the handle, even just to call `Close()`, the behavior is undefined and depends on the nature of the resource.

You can prevent a client copying objects of your R class by declaring the copy constructor and assignment operator private and not implementing them. If taking a copy is a valid action, you should declare and define the copy constructor and assignment operator, or provide another method, such as `CloneL()`, by which the resource handle is copied safely.

If your class does not need a destructor, for example because it has no cleanup code, you are under no obligation to add one. A good example of this is a T or R class, described in more detail in Chapter 1; neither type of class has need of an explicit destructor. For C classes, any compiler-generated destructor is virtual by default since the parent class, `CBase`, defines a virtual destructor. This means that, if your C class inherits from a class which defines a destructor, it is called correctly, regardless of whether you declare and define an empty destructor or allow the compiler to do it for you.

20.7 Summary

This chapter stressed the importance of defining classes clearly and comprehensively so that code is re-used rather than re-written or duplicated. It discussed a number of issues, most of which are not just specific to Symbian OS, but relate generally to C++ best practice, including:

- a comparison of the relative merits of passing and returning by value, reference or pointer
- the use of `const` where appropriate
- functional abstraction and how (not) to expose member data
- the use of enumerations to clarify the role of a function parameter and allow it to be extended in future

- the functions a compiler will generate for a class if they are not declared.

This chapter also discussed the use of `IMPORT_C` and `EXPORT_C` on Symbian OS to export API functions for use by external client code. The choice of when to export a function is relatively straightforward:

- Virtual functions should always be exported unless the class is non-derivable or where the functions are pure virtual (because there is no code to export, except in rare cases).
- Code which exports virtual functions must also export a constructor, even if it is empty, in order for the virtual function table to be created correctly. If necessary, the constructor should be specified as protected to prevent it being called directly by clients of the class.
- Inline functions should never be exported.
- Only functions that are used outside the DLL (called either directly or indirectly through a call from an inline function) should be exported.

21

Good Code Style

We all learn from our mistakes. The trick is to learn from the mistakes of others

Anon

Symbian OS is designed to work well on devices with limited resources while still providing rich functionality. To balance the demand on system resources with their relative scarcity requires program code to be of high quality; in particular, robust, efficient and compact.

You can find advice on writing good-quality C++ code in any decent textbook on the subject. There are entire books which cover aspects of good programming style and there isn't enough space in this book to go into too much detail. Instead, I've tried to focus on a few aspects that can improve your C++ code on Symbian OS. The chapter covers some of the main features of the Symbian OS coding standards and distils advice that developers working on Symbian OS have found to be most useful. While it concentrates on issues specific to Symbian OS, it also has some tips for general good programming style. I used some excellent books on C++ for background information when researching this chapter; those I recommend in particular are listed in the Bibliography.

21.1 Reduce the Size of Program Code

Space is at a premium on Symbian OS. Where possible, you should attempt to minimize the size of your program binary and its use of memory without unduly complicating your code. In previous chapters, I've already discussed some of the factors that may increase the size of your object code. For example, Chapter 5 mentioned that the use of the `_L` macro is deprecated and you should prefer to use `_LIT` for literal descriptors. The `TRAP`/`TRAPD` macros, which I discussed in Chapter 2, can also bloat program code. The following discussion describes some other factors to consider which may help you trim the size of your program binaries.

Take Care When Using `inline`

Beware of inlining functions, since you may end up bloating your program code while you are trying to speed it up. You'll find a lot more about the hazards of inlining in Scott Meyers' *Effective C++*.[1] To summarize: the `inline` directive, when obeyed by a compiler, will replace a call to the inlined function with the body of the function, thus avoiding the overhead of a function call. However, this will increase the size of your code if the function code is anything other than trivial. (If the directive is ignored things may be even worse, since the compiler may generate multiple copies of your function code, while still making function calls to it, thus bloating your binary without any of the speed benefits from forgoing the function call.) On Symbian OS, limited memory resources typically mean that the overhead of a function call is preferable to the potential code bloat from a large section of inlined code.

Furthermore, inlined code is impossible to upgrade without breaking binary compatibility. If a client uses your function, it is compiled into their code, and any change you make to it later will force them to recompile, as I described in Chapter 18.

For both reasons, stick to the rule that you should prefer a function call and return over an inlined function unless the function is trivial, such as a simple "getter" type function, shown below:

```
class TBook
    {
public:
    ...
    inline TInt Price() const {return (iPrice);};
private:
    TInt iPrice;
    };
```

Other candidates for inlining include trivial T class constructors and templated functions, particularly when using the thin template idiom discussed in Chapter 19.

Reuse Code

Code reuse has the benefit that you get to use tried and tested code. This is great if the solution fits your problem exactly, but how many times have you found that it doesn't quite do what you want? Often you may find yourself writing code from scratch or copying and reworking code from a previous solution to solve your problem. Both courses are understandable if the prior solution doesn't fit exactly, but it is wasteful of space to end up with two or more different solutions to a common problem.

---

[1] See Item 33 of *Effective C++: 50 specific ways to improve your programs and designs*, details of which can be found in the Bibliography.

Sometimes it may be possible to abstract the common code into a utility module which can be used by both solutions. Of course, this isn't always possible, since you may not be able to modify the original source. If you cannot modify the original code, do you have to copy it for your solution, or can you use it and perform additional tasks to get the result you need? If you do have to copy and modify the code, try to copy only as much as is required, and make sure you adapt it where you need to, rewriting it completely if necessary so it is efficient for the task you are trying to achieve.

When writing a class, you should endeavor to make it easy to use and thus to reuse. Sometimes code is dismissed as being impossible to reuse because it's so hard to work out what it's doing or how best to use it. This is partly a problem with the language, because C++ is complex and it is easy to use it incorrectly. Chapter 20 discusses some of the ways to define a clear and efficient API, which will make your life easier in the testing, maintenance and documentation phases and will improve the chances that potential clients will adopt it, rather than duplicate effort and code by writing their own version.

Refactor Code

Martin Fowler describes the concept of code refactoring in his book *Refactoring: Improving the Design of Existing Code*, details of which can be found in the Bibliography. As a module evolves, it may begin to lose its structure as new functions are added to enhance and extend the program. Even well-designed code, which initially performed a specific job well, becomes increasingly complex when modified to support additional requirements built on top of the original design. In theory, the program should be redesigned rather than simply extended. However, this usually doesn't happen for numerous reasons, often related to the release schedule, which may be unprepared for redesign and reimplementation of code which is apparently fit-for-purpose.

Fowler suggests that refactoring techniques can be used as a trade-off between the long-term pain of additional code complexity and the short-term pain of redesign, such as the introduction of new bugs and glitches. Refactoring doesn't usually change the functionality of existing code but, rather, changes its internal structure to make it easier to work with as it evolves and is extended. The changes come through small steps, such as renaming methods, abstracting similar code from two classes into a separate shared superclass or moving member variables between classes when ownership changes become appropriate. The idea of refactoring is not to extend the code itself; this should be a distinctly separate operation to meet new requirements. But refactoring should take place whenever code is extended or bugs are fixed; it's an ongoing process to reduce complexity and increase code reuse. You should make sure you have good test code in place when you refactor and run the tests after each modification to ensure that you do not introduce regressive bugs into the code.

Remove Debug, Test and Logging Code in Release Builds

Chapter 16 discusses when assertion statements are appropriate in release builds; otherwise, any code you have added for debugging, testing and logging should be compiled out of release code using the appropriate macros. This allows you to minimize the overhead of diagnostic checking, in terms of runtime speed and code size, in production code.

21.2 Use Heap Memory Carefully

Memory management is an important issue on Symbian OS, which provides an infrastructure to help catch memory leaks, use memory efficiently and test that code fails gracefully under low-memory conditions. I'll discuss each of these further in this section. The issue of good memory management is so crucial to writing effective code on Symbian OS that you'll find most chapters in this book touch on it to some extent. This section references those other chapters where relevant.

Prevent Memory Leaks

In Chapter 3, I described why you should use the cleanup stack and how to do so in order to handle leaves safely. The general rule to follow is that, when writing each line of code, you should ask yourself: "Can this code leave? If so, will all the currently allocated resources be cleaned up?" By ensuring that all resources are destroyed in the event of a leave, you are preventing potential memory leaks when exceptions occur. This includes using the two-phase construction idiom for C classes, because a C++ constructor should never leave. I discussed this in more detail in Chapter 4.

You should, of course, ensure that all resources are cleaned up under normal conditions too. For example, every resource an object owns should be freed, as appropriate, in a destructor or cleanup function. It's a common mistake to add a member variable to a class but to forget to add corresponding cleanup code to the destructor. I described how to check that your code doesn't leak in Chapter 17.

Be Efficient

As well as preventing leaks, you should use memory efficiently to prevent wasting the limited resources available on a typical mobile phone. For example, you may want to consider whether the object size of your heap-based classes can be reduced. Possible slimming techniques include replacing a number of boolean flags with a bitfield and checking that you are using the Symbian OS container classes efficiently.[2] You should also check that multiple copies of filenames, or other string data, are not copied unnecessarily between components in a module – perhaps passing them instead as reference parameters between components, where possible, to avoid duplication.

There are many other ways to reduce memory use, but it's also sensible to consider how complex your code may become. There may be a degree of compromise between preserving the readability of your code and shaving off some extra bytes of memory consumption. As I've already mentioned, it's important to remember that if your code is hard to understand, it is less likely to be reused. Potential clients may prefer to re-implement code rather than spend time figuring out how it works, and worrying about how they will maintain it.

By extension, you should prefer to use the classes provided by Symbian OS, rather than re-implement your own version of, for example, a string class. As I described in Chapters 5 and 6, the descriptor classes were carefully designed for memory efficiency. The Symbian OS classes have been written and maintained by those with expertise in dealing with restricted memory resources, and the classes have been extensively tried and tested as the OS has evolved. While the experience of writing your own utility classes and functions is irreplaceable, you get guaranteed efficiency and memory safety from using those provided by Symbian OS.

When allocating objects on the heap (and on the stack as described in the following section), the lifetime of the object is fundamental. On Symbian OS, programs may run for months and the user may rarely reset their phone. For this reason, objects should be cleaned up as soon as their useful lifetime ends – whether normally or because of an error condition. You should consider the lifetime of temporary objects allocated on the heap and delete them as soon as they have served their purpose. If you retain them longer than is necessary, the RAM consumption of the application may be higher than necessary.

Of course, there's often a trade-off between memory usage and speed. For example, you may decide to conserve memory by creating an object

---

[2] The `RArray`- and `CArray`-derived container classes take a granularity value on construction which is used to allocate space for the array in blocks, where the granularity indicates the number of objects in a block. Your choice of granularity must be considered carefully because, if you over-estimate the size of a block, the additional memory allocated will be wasted. For example, if a granularity of 20 is specified, but the array typically contains only five objects, then the space reserved for 15 objects will be wasted. Of course, you should also be careful not to under-estimate the granularity required, otherwise frequent reallocations will be necessary. Chapter 7 discusses the use of the Symbian OS container classes in more detail.

only when necessary, and destroying it when you've finished with it. However, if you need to use it more than once, you will reduce speed efficiency by duplicating the effort of instantiating it. On these occasions, individual circumstances will dictate whether you use the memory and cache the object or take the speed hit and create it only when you need it, destroying it afterwards. There is some middle ground, though: you may prefer not to instantiate the object until its first use and then cache it.

Once you have deleted any heap-based object, you should set the pointer to NULL[3] to prevent any attempt to use it or delete it again. However, you don't need to do this in destructor code: when member data is destroyed in a destructor, no code will be able to re-use it because it drops out of scope when the class instance is deleted.

In the following example, don't worry too much about the details of class `CSwipeCard` or the two security objects it owns, `iPassword` and `iPIN`. However, consider the case where a `CSwipeCard` object is stored on the cleanup stack and `ChangeSecurityDetailsL()` is called upon it. First the two current authentication objects are deleted, then a function is called to generate new PIN and password objects. If this function leaves, the cleanup stack will destroy the `CSwipeCard` object. As you can see from the destructor, this will call `delete` on `iPIN` and `iPassword` – consider what would happen if I had not set each of these values to NULL. The destructor would call `delete` on objects that had already been destroyed, which isn't a safe thing to do: the result is undefined, though in debug builds of Symbian OS it will certainly raise a panic (`USER 44`). Note that I've not set the pointers to NULL in the destructor – as I mentioned above, at this point the object is being destroyed, so no further access will be made to those pointers. It's safe to call `delete` on a NULL pointer, so setting the deleted pointer to NULL after deletion prevents an unnecessary panic if the method which creates the new PIN and password objects leaves.

`ChangeSecurityDetailsL()` is also called from `ConstructL()`, at which point the `iPIN` and `iPassword` pointers are NULL (because the `CBase` parent class overloads `operator new` to zero-fill the object's memory on creation, as I described in Chapter 1).

```
class CPIN;
class CPassword;

class CSwipeCard : public CBase
    {
```

[3] As an aside: in debug builds of Symbian OS, when a heap cell is freed, the contents are set to `0xDE`. If you are debugging and spot a pointer to `0xDEDEDEDE`, the contents have been deleted but the pointer has not been NULLed. Any attempt to use this pointer will result in a panic.

```
public:
    ~CSwipeCard();
    void ChangeSecurityDetailsL();
    static CSwipeCard* NewLC(TInt aUserId);
private:
    CSwipeCard(TInt aUserId);
    void ConstructL();
    void MakePasswordAndPINL();
private:
    TInt iUserId;
    CPIN* iPIN;
    CPassword* iPassword;
    };

CSwipeCard::CSwipeCard(TInt aUserId)
: iUserId(aUserId) {}

CSwipeCard::~CSwipeCard()
    {
    delete iPIN; // No need to NULL the pointers
    delete iPassword;
    }

CSwipeCard* CSwipeCard::NewLC(TInt aUserId)
    {
    CSwipeCard* me = new (ELeave) CSwipeCard(aUserId);
    CleanupStack::PushL(me);
    me->ConstructL();
    return (me);
    }

void CSwipeCard::ConstructL()
    {
    ChangeSecurityDetailsL(); // Initializes the object
    }

void CSwipeCard::ChangeSecurityDetailsL()
    {
    // Destroy original authentication object
    delete iPIN;
    // Zero the pointer to prevent accidental re-use or double deletion
    iPIN = NULL;
    delete iPassword;
    iPassword = NULL;
    MakePasswordAndPINL(); // Create a new random PIN and password
    }
```

An alternative, slightly more complex, implementation of `ChangeSecurityDetailsL()` could create the new PIN and password data in temporary descriptors before destroying the current `iPIN` and `iPassword` members. If either allocation fails, the state of the `CSwipeCard` object is retained – whereas in the implementation above, if `MakePasswordAndPINL()` leaves, the `CSwipeCard` object is left in an uninitialized state, having no password or PIN values set. The choice of implementation is usually affected by what is expected to happen if `ChangeSecurityDetailsL()` leaves.

```
void CSwipeCard::ChangeSecurityDetailsL() // Alternative implementation
    {// MakePasswordAndPINL() has been split into two for simplicity
    CPIN* tempPIN = MakePINL(); // Create a temporary PIN
    CleanupStack::PushL(tempPIN);
    CPassword* tempPassword = MakePasswordL();
    CleanupStack::PushL(tempPassword);
    delete iPIN;        // No need to NULL these, nothing can leave
    delete iPassword;   // before they are reset
    iPIN = tempPIN;
    iPassword = tempPassword;
    CleanupStack::Pop(2, tempPIN); // Owned by this (CSwipeCard) now
    }
```

Finally in this section on memory efficiency, it's worth mentioning that you should avoid allocating small objects on the heap. Each allocation is accompanied by a heap cell header which is four bytes in size, so it's clear that allocating a single integer on the heap will waste space, as well as having an associated cost in terms of speed and heap fragmentation. If you do need a number of small heap-based objects, it is best to allocate a single large block of memory and use it to store objects within an array.

Degrade Gracefully – Leave if Memory Allocation Is Unsuccessful

You should code to anticipate low memory or other exceptional conditions. To save you coding a check that each and every heap allocation you request is successful, Chapter 2 describes the overload of `operator new` which leaves if there is insufficient heap memory for the allocation. Leaves are the preferred method of coping with exceptional conditions such as low memory on Symbian OS. When a memory failure occurs and causes your code to leave, you should have a higher level `TRAP` to "catch" the failure and handle it gracefully (for example, by freeing some memory, closing the application, notifying the user, or whatever seems most appropriate).

Symbian OS runs on limited-resource mobile phones which will, at times, run out of memory. So you must prepare your code to cope under these conditions. To help you verify that your code performs safely under low memory conditions, Symbian OS provides a variety of test tools and macros that you can use in your own test code to simulate low memory conditions. I described these in detail in Chapter 17.

Check that Your Code Does not Leak Memory

I've already discussed how to prevent memory leaks, but, of course, this doesn't mean that they won't occur. It is important to test your code regularly to check that it is still leave-safe and that no leaks have been introduced by maintenance or refactoring. Symbian OS provides macros and test tools to test the heap integrity and to show up leaks when your code unloads. I describe how to use these in Chapter 17. Leaks can be difficult to track down, so it is far easier to test for them as part of the regular testing cycle, in order to catch regressions close to the time at which they are introduced.

21.3 Use Stack Memory Carefully

For a target mobile phone running Symbian OS, the program stack size defaults to 8 KB,[4] and it doesn't grow when that size is exceeded. Actually that's not quite true, because the Symbian OS emulator running on Windows will extend the program stack if the size exceeds the limit. Consequently, you may find that your application runs perfectly well on the emulator, while it overflows the stack on hardware.

So what happens when you outgrow the stack space available? Further attempts to use the stack will result in access to a memory address that's not mapped into the address space for your process. This causes an access violation that will generate an exception – at which point the kernel will panic your code with `KERN-EXEC 3`. In fact, if you try to build code where any single function tries to allocate automatic variables occupying more than 4 KB, you will receive an unresolved external `__chkstk` link error. However, this only considers static allocation and doesn't take into account recursion, variable length looping or anything else that can only be calculated at runtime.

Both the size of the program stack, which is significantly smaller than most desktop applications have available, and the fact it does not grow dynamically, make it important that your stack usage is conservative.

Manage Recursive Code Carefully

Recursive code can use significant amounts of stack memory. To ensure that your code does not overflow the stack, you should build in a limit to the depth of recursion, where this is possible. It's a good idea to minimize

---

[4] For a component running in a separate process, e.g. a server or `.exe`, you can change the stack size from the default 8 KB, by adding the following to your. `mmp` file to specify a new stack size in hexadecimal or decimal format. For example:

```
epocstacksize 0x00003000 // 12 KB
```

Alternatively, you can create a new thread and specify a chosen stack size, as described in Chapter 10. This is particularly useful if your application makes heavy use of the stack, say through recursion, and needs more space than the default stack size makes available.

If you do hit the limit, it's worth considering why your code is using so much stack space before assigning more, because consumption of extra stack resources leaves less free for other programs. There are valid reasons why you may need more stack, but you may alternatively wish to consider modifying the program architecture or employing some of the tips described here.

the size of parameters passed into recursive code and to move local variables outside the code.

Consider the Effect of Settings Data on Object Size

To minimize object size where a class has a large amount of state or settings data, you may be able to use bitfields to store it rather than 32-bit `TBool`s or enumerations (in fact, this applies equally to stack- and heap-based classes). A good example of this is the `TEntry` class which is part of `F32`, the filesystem module. A `TEntry` object represents a single directory entry on the filesystem, and keeps track of its attributes (such as whether it is marked read-only or hidden), its name and its size. The `TEntry` class is defined in `f32file.h` – it has a single `TUint` which stores the various attributes, each of which is defined in the header file as follows:

```
const TUint KEntryAttNormal=0x0000;
const TUint KEntryAttReadOnly=0x0001;
const TUint KEntryAttHidden=0x0002;
const TUint KEntryAttSystem=0x0004;
const TUint KEntryAttVolume=0x0008;
const TUint KEntryAttDir=0x0010;
const TUint KEntryAttArchive=0x0020;
...
class TEntry
    {
public:
    ...
    IMPORT_C TBool IsReadOnly() const;
    IMPORT_C TBool IsHidden() const;
    IMPORT_C TBool IsSystem() const;
    IMPORT_C TBool IsDir() const;
    IMPORT_C TBool IsArchive() const;
public:
    TUint iAtt;
    ...
    };
```

The `iAtt` attribute setting is public data, so you can perform your own checking, although the class also provides a set of functions for retrieving information about the attributes of the file. These are implemented something like the following:

```
TBool TEntry::IsReadOnly() const
    {
    return(iAtt&KEntryAttReadOnly);
    }
TBool TEntry::IsHidden() const
    {
    return(iAtt&KEntryAttHidden);
    }
```

This is all quite straightforward and transparent to client code because the function returns a boolean value, although it doesn't store the attribute as such. As I've mentioned previously, you should aim to keep your class intuitive to use and maintain. While you may be able to shave a few bytes off the class by using bitfields, you need to consider the impact this may have on the complexity of your code, and that of your clients.

Use Scoping to Minimize the Lifetime of Automatic Variables

Simple changes to the layout of code can make a significant difference to the amount of stack used. A common mistake is to declare all the objects required in a function at the beginning, as was traditional in C code.

Consider the following example, which is flawed because it declares three large objects on the stack before they are all immediately necessary.

```
void CMyClass::ConstructL(const TDesC& aDes1, const TDesC& aDes2)
    {
    TLarge object1;
    TLarge object2;
    TLarge object3;

    object1.SetupL(aDes1);
    object2.SetupL(aDes2);
    object3 = CombineObjectsL(object1, object2);
    iUsefulObject = FinalResultL(object3);
    }
```

If either of the `SetupL()` functions fails, space will have been allocated and returned to the stack unnecessarily for at least one of the large objects. More importantly, once `object1` and `object2` have been combined to generate `object3`, it is unnecessary for them to persist through the call to `FinalResultL()`, which has an unknown stack overhead. By prolonging the lifetime of these objects, the code is wasting valuable stack space, which may prevent `FinalResultL()` from obtaining the stack space it requires.

A better code layout is as follows, splitting the code into two separate functions to ensure the stack space is only used as it is required, at the expense of an extra function call. Note that, because the parameters are all passed by reference, no additional copy operations are performed to pass them to the separate function.

```
void CMyClass::DoSetupAndCombineL(const TDesC& aDes1,
        const TDesC& aDes2, TLarge& aObject)
    {
    TLarge object1;
    object1.SetupL(aDes1);
    TLarge object2;
    object2.SetupL(aDes2);
```

```
    aObject = CombineObjectsL(object1, object2);
    }

void CMyClass::ConstructL(const TDesC& aDes1, const TDesC& aDes2)
    {
    TLarge object3;
    DoSetupAndCombineL(aDes1, aDes2, object3);
    iUsefulObject = FinalResultL(object3);
    }
```

In general, it is preferable to avoid putting large objects on the stack – where possible you should consider using the heap instead. If you make large objects member variables of C classes, you can guarantee that they will always be created on the heap.

21.4 Eliminate Sub-Expressions to Maximize Code Efficiency

If you know that a particular function returns the same object or value every time you call it, say in code running in a loop, it is sensible to consider whether the compiler will be able detect it and make an appropriate optimization. If you can guarantee that the value will not change, but cannot be sure that the compiler will not continue to call the same function or re-evaluate the same expression, it is worth doing your own sub-expression elimination to generate smaller, faster code. For example:

```
void CSupermarketManager::IncreaseFruitPrices(TInt aInflationRate)
    {
    FruitSection()->UpdateApples(aInflationRate);
    FruitSection()->UpdateBananas(aInflationRate);
    ...
    FruitSection()->UpdateMangoes(aInflationRate);
    }
```

In the above example, you may know that `FruitSection()` returns the same object for every call made in the function. However, the compiler may not be able to make that assumption and will generate a separate call for each, which is wasteful in terms of space and size. The following is more efficient, because the local variable cannot change and can thus be kept in a register rather than being evaluated multiple times:

```
void CSupermarketManager::IncreaseFruitPrices(TInt aInflationRate)
    {
```

```
    CFruitSection* theFruitSection = FruitSection();
    theFruitSection->UpdateApples(aInflationRate);
    theFruitSection->UpdateBananas(aInflationRate);
    ...
    theFruitSection->UpdateMangoes(aInflationRate);
    }
```

Consider the following example of inefficient code to access and manipulate a typical container such as an array. The function adds an integer value to every positive value in the `iContainer` member of class `CExample` (where `iContainer` is a pointer to some custom container class which stores integers).

```
void CExample::Increment(TInt aValue)
    {
    for (TInt index=0;index<iContainer->Count();index++)
        {
        if ((*iContainer)[index]>0)
            (*iContainer)[index]+=aValue;
        }
    }
```

The compiler cannot assume that `iContainer` is unchanged during this function. However, if you are in a position to know that the functions you are calling on the container will not change it (that is, that `Count()` and `operator[]` do not modify the container object itself), it would improve code efficiency to take a reference to the container, since this can be stored in a register and accessed directly.

```
void CExample::Increment(TInt aValue)
    {
    CCustomContainer& theContainer = *iContainer;
    for (TInt index=0;index<theContainer.Count();index++)
        {
        if (theContainer[index]>0)
            theContainer[index]+=aValue;
        }
    }
```

Another improvement in efficiency would be to store a copy of the number of objects in the container rather than call `Count()` on each iteration of the loop. This is not assumed by the compiler, because it cannot be certain that the number of objects in `iContainer` is constant throughout the function. You, the programmer, can know if it is a safe optimization because you know that you are not adding to or removing from the array in that function.

(Rather than creating another variable to hold the count, I've simply reversed the `for` loop so it starts from the last entry in the container. This has no effect on the program logic in my example, but if you know you have to iterate through the contents of the array in a particular order, you can just create a separate variable to store the count, which is still more efficient than making multiple calls to retrieve the same value.)

```
void CExample::Increment(TInt aValue)
    {
    CCustomContainer& theContainer = *iContainer;
    for (TInt index=theContainer.Count();--index>=0;)
        {
        if (theContainer[index]>0)
            theContainer[index]+=aValue;
        }
    }
```

In the code that follows, I've added another optimization, which stores the integer value retrieved from the array from the first call to `operator[]`, rather than making two separate calls to it. This is an optional enhancement, which probably depends on the percentage of the array to which the `if` statement applies. As a general rule, I tend to store retrieved values if they are used twice or more, when I know they are constant between uses.

```
void CExample::Increment(TInt aValue)
    {
    CCustomContainer& theContainer = *iContainer;
    for (TInt index=theContainer.Count();--index>=0;)
        {
        TInt& val = theContainer[index];
        if (val>0)
            val+=aValue;
        }
    }
```

21.5 Optimize Late

The sections above offer a number of suggestions to improve the style of your code, concentrating on reducing memory usage and code size and increasing efficiency. You should bear them in mind, bringing them naturally into your code "toolbox" to make your Symbian OS code more effective.

However, a note of caution: don't spend undue amounts of time optimizing code until it is finished and you are sure it works. Of course, you should always be looking for an optimal solution from the start, using good design and writing code which implements the design while being simple to understand and maintain. A typical program spends most of its time executing only a small portion of its code. Until you have identified which sections of the code are worthy of optimization, it is not worth spending extra effort fine-tuning the code. When you have profiled your code, you will know what to optimize and can consider the tips above. Then is the time to take another look at the code and see if the general rules I've mentioned here (minimizing stack usage, releasing heap memory early and protecting it against leaks, and coding efficiently when the compiler cannot make assumptions for you) can be applied.

And finally, you may find it useful to look at the assembler listing of your code. You can generate this using the Symbian OS command line `abld` tool, the use of which will be described in detail in your SDK documentation.

`abld listing` generates an assembler listing file for the source files for a project. For a project with a single source file, `testfile.cpp`, `abld listing winscw udeb` creates an assembler listing file (`testfile.lst.WINSCW`) for the code in `testfile.cpp`. You can inspect this to find out which optimizations have been applied by the compiler, before coding the optimization yourself. The listing will also give you an idea of which code idioms are particularly "expensive" in terms of the number of instructions required.

21.6 Summary

To program effectively on Symbian OS, your code needs to be robust, efficient and compact. While many of the rules for writing high quality C++ are not specific to Symbian OS, this chapter has focused on a few guidelines which will definitely improve code which must work with limited memory and processor resources and build with a small footprint to fit onto a ROM. The chapter reviewed some of the main areas covered by the Symbian OS coding standards and described the main principles that developers working on Symbian OS adhere to, namely minimizing code size, careful memory management for both stack and heap memory and efficiency optimization.

You can find more advice on writing good C++ in the books listed in the Bibliography.

Appendix

Code Checklist

The following list is based on the internal Symbian Peer Review Checklist.

Declaration of Classes

- A class should have a clear purpose, design and API.
- An exported API class should have some private reserved declarations for future backward compatibility purposes.
- The API should provide the appropriate level of encapsulation, i.e. member data hiding.
- Friendship of a class should be kept to a minimum in order to preserve encapsulation.
- Polymorphism should be used appropriately and correctly.
- A class should not have exported inline methods.
- A non-derivable class with two-phase construction should make its constructors and `ConstructL()` private.
- `const` should be used where possible and appropriate.
- Overloaded methods should be used in preference to default parameters.
- Each C class should inherit from only one other C class and have `CBase` as the eventual base class.
- Each T class should have only stack-based member variables.
- Each M class should have no member data.

Header Files

- Header files should contain "in-source" documentation of the purpose of each class.
- The inclusion of other header files should be minimized by using forward declaration if required.
- Header files should include the correct `#ifndef...#define...#endif` directive to prevent multiple inclusion.
- The number of `IMPORT_C` tags should match the corresponding `EXPORT_C` tags in the `.cpp` file.

Comments

- Comments should match the code.
- Comments should be accurate, relevant, concise and useful.
- Each class should have "in-source" documentation of its purpose in header file.
- Each method should have "in-source" documentation of its purpose, parameters and return values in the `.cpp` file.

Constructors

- A `CBase`-derived class does not need explicit initialization.
- Member data of T and R classes should be explicitly initialized before use.
- Member data should be initialized in the initializer list as opposed to the body.
- A C++ constructor should not call a method that may leave.
- Two-phase construction should be used if construction can fail with anything other than `KErrNoMemory`.
- `NewL()` and `NewLC()` methods should wrap allocation and construction where necessary.

Destructors

- All heap-allocated member data owned by the class should be deleted.
- There should be a `NULL` pointer check before any dereference of a member pointer.
- Member pointers do not need to be set to `NULL` after deletion in the destructor.
- Member pointers deleted outside the destructor should be set to `NULL` or immediately re-assigned another value.
- A destructor should not leave.

Allocation and Deletion

- For a new expression, either use `new (ELeave)`, or check the return value against NULL if leaving is not appropriate.
- For a call to `User::Alloc()`, either use `AllocL()`, or check the return value against NULL if leaving is not appropriate.
- Objects being deleted should not be on the cleanup stack or referenced by other objects.
- C++ arrays should be de-allocated using the array deletion operator, `delete [] <array>`.
- Objects should be de-allocated using `delete <object>`.
- Other memory should be de-allocated using the method `User::Free (<memory>)`.

Cleanup Stack and Leave Safety

- If a leaving method is called during an object's lifetime, then the object should either be pushed on to the cleanup stack or be held as a member variable pointer in another class that is itself leave-safe.

- An object should not simultaneously be owned by another object and held on the cleanup stack when a leaving method is called.
- Calls to `CleanupStack::PushL()` and `CleanupStack::Pop()` should be balanced (look out for functions that leave objects on the cleanup stack, such as `NewLC()`).
- If possible, use the `CleanupStack::Pop()` and `PopAndDestroy()` methods which take pointer parameters, because they catch an unbalanced cleanup stack quickly.
- Where possible, use `CleanupStack::PopAndDestroy()` rather than `Pop()` followed by `delete` or `Close()`.
- Local R objects should be put on the cleanup stack using `CleanupClosePushL()`, usually after the object has been "opened".
- Errors caught by a `TRAP` harness that are ignored must be clearly commented and explained.
- The code within a `TRAP` harness should be able to leave.
- Error processing after a `TRAP` that ends with an unconditional `User::Leave()` with the same error code should be re-coded using the cleanup stack.
- Use the LeaveScan tool to check source code for functions that can leave but have no trailing L.

Loops and Program Flow Control

- All loops should terminate for all possible inputs.
- The bracing should be correct on `if...else` statements.
- Optimize loops by using the clearest type (do...`while` or `while`) in each case.
- There should be a `break` statement for each `case` of a `switch` statement – or a comment if a flow-through to the next `case` statement is meant.
- A `switch` statement shouldn't contain `case` statements that `return` from the function unless every `case` in that `switch` statement contains a `return`.
- A `switch` statement should be used to indicate state control. Each `case` statement should contain only a small amount of code; large chunks of code should be moved into separate functions to maintain the readability of the `switch` statement.

- There should be a `default` in every `switch` statement.
- Flow control statements should be used rather than extra boolean variables to manage program flow.

Program Logic

- Stack usage should be reasonable, i.e. don't store large objects on the stack.
- Errors should propagate properly and not be changed or remapped to e.g. `KErrGeneral`.
- Pointers should be manipulated and accessed correctly, i.e. check for invalid address access, pointer overflow, etc.
- Bitmasks should be used properly (`&` to read, | to set, ~ to reset).
- Unexpected events and inputs should be handled gracefully, i.e. use error returns, leaves or panics appropriately.
- Hard-coded "magic" values should not be used.
- Logging statements should be informative and in useful places.
- Code should be efficient, e.g. do not copy things unnecessarily, do not create too many temporaries, do not use inefficient algorithms.
- Casts should be in C++ style such as `static_cast`, `reinterpret_cast` or `const_cast` rather than in C style.
- A cast should only be used where absolutely necessary and should be the correct type for the expression.

Descriptors

- Descriptor character size should be correct, i.e. 8 bits for binary data and ASCII-like text; 16 bits for explicitly Unicode text; no suffix for normal text.
- `_L` should not be used for literal data: use `_LIT` instead (except for test code, etc.).
- Assignment to `TPtr` or `TPtrC` does not redirect the pointer; `Set()` is required for that purpose.
- `HBufC::Des()` and `TBufC::Des()` are not required to use the object as an unwritable descriptor.

- `HBufC::Des()` and `TBufC::Des()` are not required for assignment: these classes define assignment operators.
- Use `AppendNum()` or `Num()` to format a number instead of `TDes::AppendFormat()` or `Format()`.
- Use `Append()` instead of `TDes::AppendFormat()` or `Format()` to concatenate strings.
- Arguments to `TDes::AppendFormat()` or `Format()` should match the format of the string and the descriptor must be passed by pointer.
- Use descriptor functions such as `Find()` and `Locate()` rather than re-implementing this functionality.
- Iterating over a descriptor using `operator[]` is expensive; consider using C++ pointer arithmetic and the `TDesC::Ptr()` function instead.
- Descriptors passed to asynchronous operations should persist until the operation completes, e.g. do not use a stack-allocated descriptor.

Containers

- The least-derived type of container should be used in declarations.
- The initial size and granularity of the container should be appropriate.
- All owned objects in the container should be destroyed (or ownership passed) before the container is destroyed.
- The appropriate container for the purpose should be used.
- Thin templates should be used for any templated code.

Glossary

Term	Definition	Chapter
active object	Responsible for issuing requests to asynchronous service providers and handling those requests on completion. Derives from `CActive`.	8, 9
active scheduler	Responsible for scheduling events to the active objects in an event-handling program.	8, 9
asynchronous service provider	A system, component or class, which provides a service asynchronously, typically to an active object. Asynchronous requests are indicated by function calls with a `TRequestStatus` reference parameter.	8, 9
backward compatibility	Updating a component in such a way that client code which used the original version can continue to work with the updated version. See also **forward compatibility**.	18
binary compatibility	Modifying a library in such a way that code which linked against the previous version can continue to run with the new version without needing recompilation.	18
C function	In the context of Symbian OS, this refers to a function with a trailing C in its name, e.g. `NewLC()`. If successful, this type of function leaves one or more objects on the cleanup stack when it returns.	3

Term	Definition	Chapter
chunk	A unit of memory allocation, where a linear region of RAM is mapped into contiguous logical addresses.	13
clanger	The Clangers are a race of highly-civilized, small, bright pink, long-nosed, mouse-shaped persons who stand upright on big flappy feet (see ***www.clangers.co.uk*** for more information).	throughout
cleanup stack	Takes responsibility for cleaning up the objects pushed onto it if a leave occurs.	3
client interface function	A member function in a client-side session (or subsession) class that sends a specific message request to a server, identified by an opcode.	11, 12
context switch	A switch in execution between one thread or process and another. This involves saving the context of the currently executing thread or process (i.e. its address space mappings) and restoring the context of the newly executing thread or process.	10, 11, 12
descriptor	A string, so-called because it is self-describing. All descriptor classes derive from `TDesC`, which describes an area of memory used as either a string or binary data.	4, 5
descriptorize	To marshal a flat data object, typically a struct or T class, into a descriptor for inter-thread data transfer.	6, 11, 12
DLL (dynamic link library)	Dynamic link libraries contain functions that are linked and stored separately from the programs that use them.	13
DLL export table	An area in the DLL which lists the DLL addresses of exported items. On Symbian OS, these items are indexed by ordinal.	13
E32	Collective term for the base components of Symbian OS.	
ECOM	A generic framework in Symbian OS from v7.0 for use with plug-in modules.	14

Term	Definition	Chapter
emulator	An implementation of Symbian OS hosted on a PC running Microsoft Windows. Compare with **target hardware**.	13
entry point	A function called when a DLL is loaded or attached by a thread or process. In Symbian OS, the function `E32Dll()` is usually coded as the entry point.	13
EPOC, EPOC32	Earlier names for Symbian OS. Rumors that this stood for 'Electronic Piece of Cheese' are unconfirmed.	
exec call	A call to the kernel via a software interrupt (see this useful paper by John Pagonis for more details ***www.symbian.com/developer/techlib/papers/userland/userland.pdf***)	10
F32	Collective term for the components making up the Symbian OS file server.	
FEP (Front-End Processor)	Allows the input of characters by a user, for example, by handwriting recognition or voice, and converts the input transparently into text.	13
file server	The server thread that mediates all file system operations. All application programs that use files are clients of this thread.	11
flat buffer	A dynamic buffer using flat storage, i.e. a single segment of memory (see also **segmented buffer**).	7
forward compatibility	Updating a component in such a way that code which works with the new version also works with the original version. See also **backward compatibility**.	18
framework	A component that allows its functionality to be extended by plug-in modules, sometimes known as framework extensions. The framework loads the extensions as required at runtime.	13, 14
framework extension	Provides plug-in functionality to a framework, typically by implementing interfaces defined by that framework. On Symbian OS, a framework extension is implemented as a polymorphic DLL.	14

Term	Definition	Chapter
freeze	The process of fixing the association between name and ordinal, for functions exported from a DLL. It is a colloquial reference to a `.def` file (see module definition file).	13
import library (`.lib`)	A static library that an application program uses at link time to resolve references to symbols defined in a corresponding DLL. Information from this library concerning the location of export functions in the DLL is included in the application's executable file.	13
IPC (Inter-Process Communication)	Used by clients to communicate with out-of-process servers on Symbian OS.	11, 12
just-in-time (JIT)	Just-in-time debugging is used to attach a debugger to a process at the point at which it is about to terminate.	15
kernel	The core of the operating system. It manages memory, loads processes and libraries and schedules threads for execution.	10
L function	A function with a trailing L in its name, e.g. `NewL()`, which indicates that it may leave.	2
leave	Symbian OS exception handling. A function leaves when `User::Leave()` is called. It causes a return to the current trap harness.	2
mixin	An interface class for "mixing in" with primary base classes, and the only form of multiple inheritance encouraged on Symbian OS.	1
module definition file (`.def`)	A file that specifies an ordinal value for each named function or data item to be exported from a DLL. It is used to insert ordinal information into the DLL export table, allowing exports to be accessed by ordinal from clients of the DLL.	13, 18
null thread	The lowest-priority thread in the system, belonging to the kernel. It only runs when there is no higher priority thread to run and usually puts the phone into power-saving mode.	10

Term	Definition	Chapter
OOM (out-of-memory)	An error caused when the available RAM has been exhausted, preventing further heap allocation.	throughout
panic	A run-time exception on Symbian OS that terminates the current thread. Panics tend to be caused when assertion statements fail.	15, 16
plug-in	A polymorphic interface DLL used to extend a framework application.	13, 14
polymorphic DLL	A DLL that provides plug-in functionality to another program by implementing a defined interface. These are typically known as plug-ins and are loaded dynamically by another program (usually a framework) using `RLibrary::Load()`. Compare with **shared library**.	13, 14
process	The unit of memory protection on Symbian OS. One user process may not access another's memory. A process may contain one or more threads.	10
recognizer	A plug-in that examines data from a file and returns its data (MIME) type, if recognized.	13
request semaphore	A semaphore associated with a thread, which is used to indicate the completion of an asynchronous request.	8, 9
request status	An object that indicates the completion status of a request, represented by `TRequestStatus`.	8, 9
ROM	Read-Only Memory. This is permanent memory that can be read but not (easily) written to. It holds code and data that must persist when the phone is switched off. On Symbian OS, it is usually identified as the Z: drive.	13
SDK	Software Development Kit.	
segmented buffer	A dynamic buffer using segmented storage, which consists of a doubly-linked list of segments. Segments are added when the buffer needs to expand and removed when it is compressed. See also **flat buffer**.	7

Term	Definition	Chapter
server call	A call to the kernel in which the kernel server thread runs on behalf of the user program. See this useful paper by John Pagonis for more information ***www.symbian.com/developer/techlib/papers/userland/userland.pdf***	11, 12
session	The channel of communication between a client and a server.	11, 12
shared library	A library that is loaded by the DLL loader when an executable that links to it is loaded. Compare with **polymorphic DLL**.	13
static library	Other executable code may link to this library to resolve references to exported functions. Builds with a `.lib` extension.	13
target hardware	A phone handset running Symbian OS (compare with **emulator**)	10, 13
thin template	An idiom used to minimize code bloat from the use of standard C++ templates.	19
thread-local storage (TLS)	A machine word of memory that may be used to anchor information in the context of a DLL or a thread. Used instead of writable static data, which is not supported for Symbian OS DLLs.	13
`TRAP, TRAPD`	Trap harness macros within which leaving code may be executed and any leaves "caught". Can be likened to a combination of `try` and `catch` in standard C++.	2
two-phase construction	An idiom used on Symbian OS to ensure that an object can be initialized safely using leaving code.	4
UID (unique identifier)	A unique 32-bit number used in a compound identifier to identify an object, file type, etc. When users refer to "UID" they often mean UID3, the identifier for a particular program.	13
UID type	A set of three UIDs that, in combination, identify a Symbian OS object.	13

Term	Definition	Chapter
UID1	The first UID in a compound identifier (UID type). It identifies the general type of a Symbian OS object and can be thought of as a system-level identifier; for example executables, DLLs and file stores are all distinguished by UID1.	13
UID2	The second UID in a compound identifier (UID type). It is used to distinguish within a type (e.g. between types of DLL) and can be thought of as an interface identifier. For example, static interface and polymorphic interface DLLs are distinguished by UID2.	13
UID3	The third UID in a compound identifier (UID type). It identifies a particular subtype and can be thought of as a project identifier (for example, UID3 may be shared by all objects belonging to a given program, including library DLLs, framework DLLs, and all documents).	13
Unicode	A 16-bit character set that is used to assign unique character codes for a wide range of languages.	5, 6

Bibliography and Online Resources

Books on Symbian OS

Richard Harrison (2003)
Symbian OS C++ for Mobile Phones
Publisher: John Wiley and Sons Ltd
ISBN: 0470856114

Richard Harrison (2004)
Symbian OS C++ for Mobile Phones Vol 2
Publisher: John Wiley and Sons Ltd
ISBN 0470871083

Martin Tasker, Jonathan Allin, Jonathan Dixon, Mark Shackman, Tim Richardson and John Forrest (2000)
Professional Symbian Programming: Mobile Solutions on the EPOC Platform
Publisher: Wrox Press
ISBN: 186100303X

Leigh Edwards and Richard Barker (2004)
Developing Series 60 Applications: A Guide for Symbian OS C++ Developers
Publisher: Addison Wesley
ISBN: 0321227220

J. Jipping (2002)
Symbian OS Communications Programming
Publisher: John Wiley and Sons Ltd
ISBN: 0470849487

Books on C++

Martin Fowler (1999)
Refactoring: Improving the Design of Existing Code
Publisher: Addison Wesley
ISBN: 0201485672

Martin Fowler and Kendall Scott (2003)
UML Distilled: A Brief Guide to the Standard Object Modeling
Publisher: Addison Wesley
ISBN: 0321193687

Stanley Lippman (1996)
Inside the C++ Object Model
Publisher: Addison Wesley
ISBN: 0201834545

Scott Meyers (1997)
Effective C++: 50 Specific Ways to Improve Your Programs and Designs, 2nd Edition
Publisher: Addison Wesley
ISBN: 0201924889

Scott Meyers (1996)
More Effective C++: 35 New Ways to Improve Your Programs and Designs
Publisher: Addison Wesley
ISBN: 020163371X

Bjarne Stroustrup (2000)
The C++ Programming Language, Special Edition
Publisher: Addison Wesley
ISBN: 0201700735

Herb Sutter (1999)
Exceptional C++
Publisher: Addison Wesley
ISBN: 0201615622

Symbian OS on the Internet

Symbian Press Website
www.symbian.com/books

Symbian Developer Network
www.symbian.com/developer

Symbian Developer Network Support Forums
www.symbian.com/developer/public/index.html

Jo Stichbury's website
www.WhoShavesTheBarber.com

Sony Ericsson Developer World
http://developer.sonyericsson.com

Forum Nokia
http://forum.nokia.com/main.html

Index

Printed in the United States
92638LV00004B/45/A

9 780470 021309